**Energy Storage for Power
System Planning and
Operation**

Energy Storage for Power System Planning and Operation

Zechun Hu
Department of Electrical Engineering
Tsinghua University
China

This edition first published 2020

Registered Offices
John Wiley & Sons, Inc., 111 River Street, Hoboken, NJ 07030, USA
John Wiley & Sons Singapore Pte. Ltd, 1 Fusionopolis Walk, #07-01 Solaris South Tower, Singapore 138628

Editorial Office
The Atrium, Southern Gate, Chichester, West Sussex, PO19 8SQ, UK

For details of our global editorial offices, customer services, and more information about Wiley products visit us at www.wiley.com.

Wiley also publishes its books in a variety of electronic formats and by print-on-demand. Some content that appears in standard print versions of this book may not be available in other formats.

Library of Congress Cataloging-in-Publication data applied for

HB ISBN: 9781119189084

Cover Design: Wiley
Cover Images: Abstract background © Dmitriy Rybin/Shutterstock, Battery energy storage © Petmal/Getty Images

Set in 9.5/12.5pt STIXTwoText by SPi Global, Pondicherry, India

Printed in Singapore by Markono Print Media Pte Ltd

10 9 8 7 6 5 4 3 2 1

Contents

Preface

The installed capacity of renewable energy generation (REG), represented by wind power and photovoltaic power generation, has been growing rapidly, changing the generation mix of traditional power systems. REG can be connected to the transmission network in a centralized manner, or can be connected to the distribution network in the form of distributed generation, thereby changing the operating rules and power flow direction of the power networks. Unlike traditional thermal power or hydropower units, wind and photovoltaic power outputs are subject to weather conditions, which are random, intermittent, and difficult to predict precisely. In order to cope with the challenges brought by the large-scale REG integration to the planning and operation of power systems, the deployment of energy storage system (ESS) has become an important and even essential solution.

At present, pumped hydroelectric storage (PHS) is the largest and most mature energy storage type applied in power systems. The optimal planning and operation methods for PHS power plants are quite mature. However, the PHS power plant has a long construction period and a large investment scale, and its development is constrained by multiple factors such as geographical conditions, land occupation, and environmental impacts. With the advancement of new energy storage technologies, e.g. chemical batteries and flywheels, in recent years, they have been applied in power systems and their total installed capacity is increasing very fast. The large-scale development of REG and the application of new ESSs in power system are the two backgrounds of this book.

In Chapter 1, energy storage technologies and their applications in power systems are briefly introduced. In Chapter 2, based on the operating principles of three types of energy storage technologies, i.e. PHS, compressed air energy storage and battery energy storage, the mathematical models for optimal planning and scheduling of them are explained. Then, a generic steady state model of ESS is derived.

Chapters 3 and 4 of this book mainly discuss the joint scheduling and bidding of ESS with REG. It is assumed that ESS and REG form a union to maximize their total revenue from electricity markets. The key challenge for the union is the randomness of the REG power output. Chapter 3 describes day-ahead scheduling models and bidding strategies of a REG-ESS union in the energy market, with a focus on the methods to deal with the random power output of REG. In Chapter 4, the scheduling and bidding strategies of a REG-ESS union are further discussed. A combined day-ahead bidding and real-time operating strategy based on linear decision rules is presented, which considers the revenues and costs of the REG-ESS union both at day-ahead and real-time stages. Forecast error of wind and solar power decreases distinctly with the reducing span of time horizon. The intraday energy markets have been built in the electricity markets of some countries/regions. In Sections 4.4 and 4.5, a rolling optimization strategy for the REG-ESS union, which considers the day-ahead and intraday biddings and real-time operational optimization decisions, is explained.

In Chapters 5–7, this book discusses three areas of ESS participating in optimal dispatch and control of power systems from different time scales, namely unit commitment (UC), optimal power flow (OPF) and automatic generation control (AGC). Energy balance is the key issue for dispatching or controlling an ESS for the optimal operation of power system, which means that the energy coupling constraints of an ESS should be carefully taken into account. For the UC problem, it is quite simple to consider the energy coupling constraints of an ESS. While for the OPF problem, the operating conditions under multiple successive time intervals, rather than a single time period, should be considered and optimized. Chapter 5 first introduces a deterministic UC formulation, which considers the optimal scheduling of ESS. Then, a scenario-based and a robust security constrained UC formulations are derived. In Chapter 6, two types of solutions for the multi-period OPF problem are explained, namely an interior point method and a semi-definite programming method. Interior point methods are popular for solving the single-period OPF problem, and methods based on semi-definite programming are research hotspots in recent years. The semi-definite programming method introduced in Chapter 6 is relatively basic, and readers can learn more comprehensive research results and in-depth discussions following the references. In Chapter 7, this book introduces the methods of frequency regulation simulation, control and capacity quantification considering the participation of ESS in secondary frequency regulation (i.e., AGC). Compared with the thermal generating units, a battery or flywheel energy storage system can respond to frequency regulation commands sent from power system control center with shorter delay and faster ramping speed. Therefore, technical problems on how to optimally make use of ESS for AGC are discussed, including the optimal allocation of frequency regulation commands to ESS and the required capacities for AGC contributed from both ESS and thermal generating units.

Chapters 8 and 9 of this book discuss the optimal planning of ESS connected to transmission network and distribution network, respectively. For the transmission expansion planning problem considering ESS deployment, the decision variables include the locations, power and energy capacities of ESSs, which greatly increase the difficulties of modeling and solving the problem. A joint planning model for transmission network expansion and ESS deployment is derived in Section 8.3, which considers the active power loss costs. In Section 8.4, the mathematical formulation for transmission expansion and ESS deployment planning considering the daily operating conditions is explained in order to evaluate the energy capacity of ESS more precisely. Chapter 9 introduces the optimal planning models and solution methods for deploying ESSs in distribution network with distributed generations (DGs). By transforming the nonlinear power flow constraints of a distribution network into second-order cone constraints, the ESS planning problem is formulated as a mixed integer second-order cone programming (SOCP) problem. In order to consider the uncertain power outputs of DGs and load variations, multiple typical days are selected to represent the changing operating states within a whole year. For large-scale distribution networks, the established mixed-integer SOCP problem is difficult to solve directly. In Section 9.4, a solution method based on generalized Benders decomposition is derived. Section 9.5 briefly discusses the optimization model and solution method for the joint planning of distribution network and ESS. This book does not specifically describe the optimal operation problem for distribution network with ESS. However, the sub-problem on the daily operation optimization of ESS in Section 9.4 can be used as a basic model for this problem.

In summary, this book focuses on the joint operation of REG and ESS, optimal operation of power system with ESS, and optimal planning of ESSs for the power networks. This book can be used as a reference book for graduate students and researchers who are interested in operation and planning of power systems. It should also be useful for technicians in power network planning, power system dispatch, and energy storage investment/operation companies. The ESS technologies and their application in power systems is developing rapidly, and more advanced research results are being and will be published. This book can be served as a basis for understanding the relevant technologies and tracking the latest research achievements.

Acknowledgements

The content of this book is mainly from the research work I carried out with the postgraduates of our research lab in Tsinghua University. The research work was supported in part by National Natural Science Foundation of China (51107060 and 51477082) and also in part by National Key Research and Development Program (2016YFB0900500).

I would like to specially thank Professor Yonghua Song for his support. I must also thank Huajie Ding, Fang Zhang, Shu Zhang, Haocheng Luo, Ge Gao and Zhe Lin for their valuable contributions. Thanks to Professor Pierre Pinson for the contributions on optimization methods.

To my family for their selfless and constant support to me.

Abbreviation List

AC	Alternating current
ACE	Area control error
ADMM	Alternating direction method of multipliers
AGC	Automatic generation control
BESS	Battery energy storage system
CAES	Compressed air energy storage
CG	Conventional generator
CSP	Concentrating solar power
CVaR	Conditional value at risk
DC	Direct current
DESS	Distributed energy storage system
DG	Distributed generation
DNO	Distribution network operator
DNP	Distribution network planning
DoD	Depth of discharge
EMS	Energy management system
ES	Energy storage
ESS	Energy storage system
FES	Flywheel energy storage
GBD	Generalized Benders decomposition
IPM	Interior point method
LHS	Latent heat storage
MILP	Mixed-integer linear programming
OPF	Optimal power flow
PHS	Pumped hydroelectric storage
PHSP	Pumped hydroelectric storage plant
PV	Photovoltaic
REG	Renewable energy generation
SCADA	Supervisory control and data acquisition

SCES	Supercapacitor energy storage
SCUC	Security-constrained unit commitment
SDP	Semidefinite programming
SFC	Secondary frequency control
SHS	Sensible heat storage
SMES	Superconducting magnetic energy storage
SO	System operator
SoC	State of charge
SOCP	Second-order cone programming
TEP	Transmission expansion planning
TES	Thermal energy storage
UC	Unit commitment
UPS	Uninterruptible power supply
V2G	Vehicle-to-grid
VaR	Value at risk
VRB	Vanadium redox battery
WF	Wind farm

1

Introduction

1.1 Evolution of Power System and Demand of Energy Storage

The normal operation of a power system constantly requires a balance of generation and demand. In traditional bulk power systems, the majority of power generation units are thermal (fossil fuel), hydro, and nuclear generators. Typically, the thermal or hydro generators are optimally dispatched to meet varying demands. Because of the huge demand for electricity to support the operation of modern society, thermal generators with a very large total installed capacity burn vast amounts of fossil resources each year. Over time, their greenhouse gas and other pollutant emissions lead to serious environmental problems.

To reduce the dependence on fossil energy, renewable energy generation (REG), represented by wind power and photovoltaic (PV) power generation, has been growing rapidly all over the world in recent years. According to data from Global Wind Energy Council (GWEC), the total installed capacity of wind worldwide was about 24 GW at the end of 2001, while this capacity reached 591 GW in 2018 [1], which increased by about 24 times. The global cumulative installed wind energy capacities from 2011 to 2018 are illustrated in Figure 1.1. In addition, as of the end of 2018, the total installed capacity of PV generation worldwide reached about 503 GW according to the statistics provided by the IEA Photovoltaic Power System Programme [2]. The total installed capacity of solar power generation was about 99.9 GW in 2018, which is comparable to all the installed capacity (about 100.9 GW) by 2012. In only 10 years, the world's total PV capacity increased by over 3600% – from 14.5 GW in 2008. The boom of PV capacity can be seen from Figure 1.2 [2, 3]. It is expected that the total installed capacity of PV will reach 1.0 TW by 2022.

With the large-scale integration of renewable energy generation into power systems, the power generation mix is gradually changing, which leads to the

Energy Storage for Power System Planning and Operation, First Edition. Zechun Hu.
© 2020 John Wiley & Sons Singapore Pte. Ltd.
Published 2020 by John Wiley & Sons Singapore Pte. Ltd.

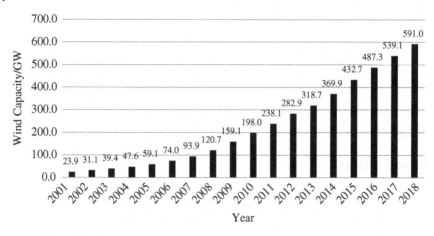

Figure 1.1 Global cumulative installed wind energy capacity.

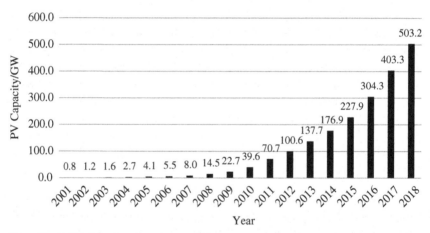

*Note: The installed PV capacity at the end of 2018 is an estimated value.

Figure 1.2 Global cumulative installed PV capacity.

changes of power flow among the power networks. Different from the traditional generation units, the power outputs of wind turbines and PV panels are constrained by weather conditions, which are difficult to forecast precisely. The random and fluctuating nature of wind and solar power brings new challenges to both

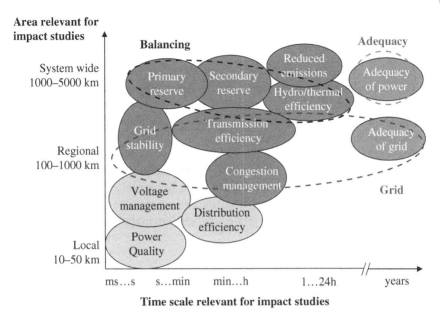

Figure 1.3 Impacts of wind power on power systems, displayed by time and spatial scales relevant for the studies [4].

planning and operation of power systems. It is more difficult to guarantee the balance between generation and demand in real-time power system operation than before.

According to different time-scales and geographical ranges, reference [4] classifies the impacts of wind power integration into three aspects: power balance, power adequacy, and network adequacy (as illustrated in Figure 1.3). Actually, the impacts of large-scale integration of PV generation are similar to that of wind power, except that PV panels cannot generate electricity during nighttime.

Reference [5] summarizes the four impacts of variable renewable generators from the aspect of power system flexibility. The first is the increased need for frequency regulation, because wind and solar power can increase the short-term variability of the net load (load minus the power of REG). The second is the increase in the ramping rate, or the speed at which load-following units must increase and decrease their output. The third impact is the uncertainty of the renewable resource. The final impact is the increase in overall ramping range and the associated reduction in minimum load, which can force baseload generators to reduce output. In summary, the increased variability and uncertainty of the net load requires a greater amount of flexibility and operating reserves in the power system.

In considering the operation of electricity markets with a high penetration of REG, three key characteristics of renewable technologies are identified in reference [6]:

1) Variability and uncertainty. The variability is partially predictable but also partially uncertain.
2) Low short-run marginal costs (SRMC). Many renewable technologies have very low operating costs, or SRMC.
3) Nonsynchronous. Wind turbines and PV panels are nonsynchronously connected to the power grid. This means that they interact with the grid in a different manner when compared to synchronously connected generators. In the absence of sophisticated power control, they do not contribute to frequency stability, system inertia, and other grid-related services in the same way as the synchronous generators.

Electricity markets that incorporate large quantities of variable renewable generation need to ensure adequate access to flexibility. Furthermore, adequate grid codes and regulatory frameworks are required to ensure appropriate provision of system stability requirements [6].

Large-scale integration of REG brings significant influence on the dynamic response characteristics and stability of a power system after a disturbance. Through a simplified model, reference [7] evaluates the impact of inertia reduction due to wind and PV power electronics on frequency stability dynamics in the continental Europe interconnected system. REG also affects the small signal stability of power systems. In reference [8], the impacts of both utility scale PVs and rooftop PVs on the small signal stability of a large power system are investigated, based on both eigenvalue analysis and transient analysis. The stability issues that are also influenced by REG include frequency, voltage, and transient stability, which are discussed in detail in reference [9]. It should be noted that modern wind turbines can provide a fast frequency (virtual inertial) response with its own characteristics. Due to the fast response time of wind farm controllers and the energy stored in wind turbine rotors, it is technically feasible to provide a rapid, but temporary, power injection.

Traditional modeling and optimization methods for power system planning should also be upgraded to handle REG. In terms of generation expansion planning, sufficient flexibility and reserve capacity should be considered to deal with the variations of renewable power. Thus, more detailed production simulations and verifications should be conducted before making the final planning decision. To consider the requirements of a flexible generation mix, a new generation expansion planning method is proposed in reference [10], while unit commitment (UC) is embedded to include the operating constraints. In reference

[11], a "high-resolution" modeling framework is proposed to optimize the generation investment in the long term, which takes into account the hourly dynamics of electricity supply and demand.

In terms of transmission network expansion planning, the objective is generally to minimize the total investment cost, while the requirements of securely transmitting power from generators to loads are satisfied [12]. Active power losses and reactive power flow are considered in some studies [13]. Under the environment of the electricity market, the competition of different market participants leads to more volatile power flow among the networks. Thus, transmission expansion planning needs to consider multiple objectives and more operational scenarios [14, 15]. Large-scale integration of renewable generation brings higher uncertainty to this problem. Reference [16] analyzes the new challenges to transmission expansion planning. A robust transmission expansion planning method is proposed in reference [17] to consider the uncertain renewable generation and loads. In reference [18], Benders decomposition algorithm is used in conjunction with Monte Carlo simulation to solve the probabilistic transmission expansion planning problem, which considers power injection and absorption uncertainties.

Distributed generation (DG) is expected to become more important in the future power system [19]. More and more integration of DGs, including PV panels and small wind turbines, are changing the power flow directions and the basic function of distribution systems [20]. Many research papers have been published on the optimal sizing and siting of DGs within a distribution system or a microgrid [21, 22]. In references [23] and [24], a multi-stage model is built for distribution network expansion planning with DGs. In reference [25], joint expansion planning of DGs and distribution network is further studied.

For a bulk power system, with the integration of renewable generation in the form of both utility scale plants and small-sized DGs, the uncertain power injections may result in reliability and stability problems, while at the distribution system level, the integration of distributed PV and wind turbines may lead to overloading or overvoltage problems. Energy storage (ES) is being widely regarded as one of the most important solutions to deal with the variations of renewable generation for its ability to add flexibility, control intermittence, and provide back-up generation to electrical networks [26–28]. Reference [28] provides a comprehensive analysis of ES leading technologies' main assets, research issues, economic benefits, and technical applications, and gives special emphasis to ES on islanded power systems. In reference [29], the prospects of several candidate storage technologies are evaluated for global energy sustainability. Reference [30] studies the problem of optimal siting and sizing of energy storage systems (ESSs) in a future European power system with 100% renewable energy generation.

1.2 Energy Storage Technologies and Their Applications in Power Systems

1.2.1 Energy Storage Technologies

ES works by moving energy through time and electricity storage is one form of ES [31]. Electrical energy can be converted to many different forms for storage, including chemical energy, mechanical energy, thermal energy, electromagnetic energy, etc. According to the form that energy is stored by the ES equipment, the mainstream ES technologies can be divided into four categories, i.e. electromagnetic ES, mechanical ES, chemical ES, and thermal energy storage (TES). Figure 1.4 illustrates the classification of ES technologies. The full names of the abbreviations are listed in Table 1.1.

1.2.1.1 Pumped Hydroelectric Storage
Pumped hydroelectric storage (PHS) is the most widely implemented large-scale ESS at present. It stores potential energy from height differences in water levels. A PHS normally consists of (i) two reservoirs located at different elevations, (ii) unit(s) to pump water to a high elevation (to store electricity in the form of hydraulic potential energy), and (iii) turbine(s) to generate electricity with the

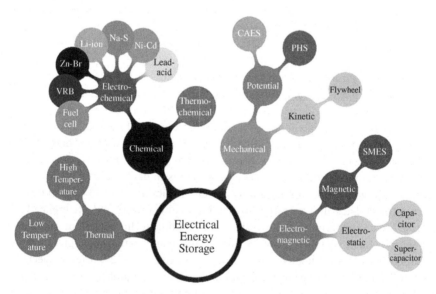

Figure 1.4 Classification of different energy storage technologies based on energy form.

Table 1.1 Abbreviations of energy storage technologies.

Abbreviation	Full name	Abbreviation	Full name
CAES	Compressed air energy storage	PHS	Pumped hydroelectric storage
Li-ion	Lithium-ion (battery)	SMES	Superconducting magnetic energy storage
Na-S	Sodium–sulfur (battery)	VRB	Vanadium redox battery
Ni-Cd	Nickel–cadmium (battery)	Zn-Br	Zinc–bromine (battery)

water returning to the low elevation (converting the potential energy to electricity when needed) [32]. Typically, reversible turbine/generator assemblies act as a pump or turbine, as necessary.

Currently, PHS is the most mature large-scale ES technology with the longest application history. PHS is popular because it can provide relatively high efficiency, large power capacity, large energy capacity, and a long life at a low cycle cost [33]. By the beginning of 2019, the total installed capacity of PHS power plants reaches 183 GW, approximately 97% of the total installed capacity of electric ES all over the world [34]. Within the European Union countries, the installed capacity of PHS accounts for about 27% of the total hydro power capacity [35].

The capacity of a PHS power plant is comparable to an average sized hydropower plant. The Bath County Pumped Storage Station in Virginia of America has a total installed capacity of 3003 MW, which began to operate from 1985. It is currently the largest PHS power plant in the world [36]. The stored capacity of PHS is only limited by the reservoir volumes. The efficiency of PHS is about 70–85% and the life time is usually more than 40 years.

The major drawback of PHS lies in the scarcity of available sites for two large reservoirs and one or two dams. A long lead time (typically ∼10 years), a high cost (typically hundreds to thousands of million US dollars) for construction, and environmental issues (e.g. removing trees and vegetation from the large area of land prior to the reservoir being flooded) are the other three major constraints in the deployment of PHS [37].

There are mainly two types of unconventional PHS: seawater PHS power plant (one is located at the northern coast of Okinawa Island, Japan, which began to operate from 1999 [38]) and underground PHS using abandoned quarries, mines, or excavated caverns as the lower reservoir [39, 40].

1.2.1.2 Compressed Air Energy Storage

Compressed air energy storage (CAES) plants are largely equivalent to pumped storage plants in terms of their applications, output, and storage capacity. However, instead of pumping water from a lower to an upper reservoir during periods of storage, in a CAES plant, ambient air is compressed and stored under pressure in either an underground structure or an above-ground system of vessels or pipes. When electricity is required, the pressurized air is heated and expanded in an expansion turbine driving a generator for producing electricity.

The first large-scale CAES plant is located in Huntorf, Germany, which has been in operation since 1978. The unit has a cavern of $\sim$310 000 m^3, converted from a solution mined salt dome located $\sim$600 m underground, coupled with 60 MW compressors providing a maximum pressure of 10 MPa. It runs on a daily cycle with eight hours of charging and can generate 290 MW for two hours [41]. The second commercialized large-scale CAES plant was built in 1991, in McIntosh, Alabama, USA. The unit compresses air to up to $\sim$7.5 MPa in an underground cavern of a solution mined salt dome 450 m below the surface. This 110 MW plant can run with its full power output for 26 hours [35]. The McIntosh CAES plant utilizes a recuperator to reuse the heat from the gas turbine, which reduces fuel consumption by $\sim$25% in comparison with the Huntorf CAES plant.

It should be noted that the round-trip efficiencies of the two CAES plants are only about 42 and 54% [42], respectively, lower than that of the PHS plant. As a result, some improved CAES systems are proposed or are under investigation. The main criterion for classification of CAES is how heat is handled during compression and prior to expansion of the air. In reference [43], CAES technologies are differentiated into diabatic, adiabatic, and isothermal concepts. The round-trip efficiency of adiabatic CAES can reach 70% [44], which is comparable to that of PHS.

Similar to the PHS, the major barrier to build a large-scale CAES plant is also the reliance on geography. Usually, the favorite sites for the CAES are rock mines, salt caverns, aquifers, or depleted gas fields.

1.2.1.3 Flywheel Energy Storage

Flywheel energy storage (FES) stores the kinetic energy in a high-speed rotating disk connected to the shaft of an electric machine and releases the stored energy when necessary. The amount of stored energy is decided by the form, mass, and rotating speed of the flywheel. There are two basic type of flywheels according to the material in the rotor. The first type uses a rotor made of advanced composite material such as carbon-fiber or graphite. These materials have very high strength to weight ratios, which enables flywheels to have the potential of high specific energy. The second class of flywheel uses steel as the main structural material in the rotor. This class includes not only traditional flywheel designs, which have large diameters, slow rotating speed, and low power and energy densities, but also some newer high-performance flywheels as well [45]. The spinning rotor within an FES device must be supported by bearings. Mechanical bearings were the first types considered, while the magnetic bearing appeared in the 1980s [46]. Recently, high-temperature superconducting bearings were used in FES.

Flywheels have good features regarding high efficiency (can be higher than 90% at rated power), long cycling life, and high power and energy densities. The fast response speed and high power density of FES makes it suitable for applications in an uninterruptible power supply (UPS), power quality improvement, power smoothing for REG, and frequency regulation for bulk power systems [47, 48]. FES can also be deployed along with an electric vehicle fast charging station to control the power interchange of the station with the power grid [49]. The large-scale FES stations with up to 20-MW-rated capacity have been built to participate in frequency regulation in the USA [50].

1.2.1.4 Lead–Acid Batteries

Lead–acid batteries are the oldest type of rechargeable batteries and are based on chemical reactions involving lead dioxide (which forms the cathode electrode), lead (which forms the anode electrode), and sulfuric acid (which acts as the electrolyte). Lead–acid batteries have high energy efficiencies and high reliability, are easy to install, and require a relatively low level of maintenance and low investment cost. In addition, the self-discharge rates for this type of battery are very low, which makes them ideal for long-term storage applications. However, the limiting factors for lead–acid batteries are the relatively low cycle life, low battery operational lifetime, and low energy density. The cycle life is negatively affected by the depth of discharge (DoD) and temperature [51–53]. The largest lead–acid battery storage station was put into commission mainly for load leveling in Chino, California, in July of 1988. It was composed of 8256 cells with $10 \, \text{MW} \times 4 \, \text{hours}$ total capacity. This station was operated for nine years until 1997 [54, 55].

New lead–acid ES technologies can be divided into two types: lead–acid carbon technologies and advanced lead–acid technologies [31, 56]. Lead–acid carbon technologies use a fundamentally different approach to lead–acid batteries through the inclusion of carbon, in one form or another, to improve the power characteristics of the battery, and to mitigate the effects of partial state of charge. Advanced lead–acid batteries are conventional valve-regulated lead–acid batteries with technologies that address the shortcomings of previous lead–acid products through incremental changes in the technology. In 2013, an advanced lead–acid battery ESS with a capacity of 36 MW/24 MWh was installed to optimally dispatch energy production from the Notrees Wind Farm in Texas [57]. Reference [58] presents a summary of advances in lead–acid batteries and the deployment of ESSs using advanced lead–acid batteries for renewable-energy and grid applications.

1.2.1.5 Nickel–Cadmium Batteries

A nickel–cadmium (Ni–Cd) battery uses nickel hydroxide and metallic cadmium as the two electrodes and an aqueous alkali solution as the electrolyte. Compared to lead–acid batteries, Ni–Cd batteries have a higher energy density and a longer lifetime [37, 51, 59]. The weaknesses of Ni–Cd batteries are: cadmium and nickel are toxic heavy metals, resulting in environmental hazards, and the battery suffers from the memory effect – the maximum capacity can be dramatically decreased if the battery is repeatedly recharged after being only partially discharged.

To date, there have been very few commercial successes using Ni–Cd batteries for utility-scale ES applications. One example is the battery storage station commissioned in 2003 at Golden Valley, Alaska, USA. This battery storage station has a rated power of 27 MW × 15 minutes with a round-trip efficiency of 72–78%, and can provide 40 MW for 7 minutes. It is mainly used for spinning reserves and grid stabilization [60, 61].

1.2.1.6 Sodium–Sulfur Batteries

A sodium–sulfur (Na–S for short) battery consists of liquid (molten) sulfur (S) at the positive electrode and liquid (molten) sodium (Na) at the negative electrode as active materials separated by a solid beta alumina ceramic electrolyte [62]. Na–S battery cells are usually designed in a tubular manner where the sodium is normally contained in an interior cavity formed by the electrolyte. Ceramic Beta-Al_2O_3 acts as both the electrolyte and the separator simultaneously. An important feature of this type of battery is its high-temperature operation, at around 300 °C.

Na–S battery cells are efficient (75 ~ 90%) and have a pulse power capability over six times their continuous rating (for 30 seconds). This attribute enables Na–S batteries to be economically used in combined power quality and peak shaving applications. Na–S batteries are environmentally benign with no emissions during operation and about 99% of the overall weight of the battery materials can be recycled [51, 63]. The major drawback is that a heat source is required, which uses the battery's own stored energy, partially reducing the battery performance, as the Na–S battery needs to operate at a high temperature [26, 35].

There are a number of Na–S battery storage stations in operation all over the world. In 2008, a station with 34 MW × 7.2 hours of capacity was installed in northern Japan for stabilizing a 51 MW wind farm [64]. Larger scale Na–S battery storage systems (several hundreds of megawatts) have been built in the United Arab Emirates for grid stabilization, frequency regulation, voltage support, power quality, load shifting, and energy arbitrage [61].

1.2.1.7 Lithium-ion Batteries

The operation of lithium-ion (Li-ion) batteries is based on the electrochemical reactions between positive lithium ions (Li+) with anodic and cathodic active materials. The cells of Li-ion batteries are made of anodic and cathodic plates, filled with liquid electrolyte material [65]. The first commercial lithium ion batteries were produced by Sony in the early 1990s [51]. Since then, improved material developments have led to vast improvements in terms of the energy density (increased from 75 to 200 Wh kg^{-1}) and cycle life (increased to as high as 10 000 cycles). Li-ion batteries also have high cycle efficiencies, up to ~97%.

The main drawbacks are that the cycle DoD can affect the Li-ion battery's lifetime. In addition, maintaining a safe voltage and operation temperature ranges are essential aspects for this technology, due to its fragility. Thus, protection circuits are required. In addition, the use of flammable organic electrolytes raises issues about security and greenness [26, 51, 52].

Lithium-ion batteries, which have achieved significant penetration into the portable consumer electronics markets, are now making the transition into electric vehicle and grid storage applications. In recent years, among the new installed ES stations, the number of projects using lithium-ion batteries is ranked number one among all the electric ES projects [34].

1.2.1.8 Vanadium Redox Batteries

The vanadium redox battery (VRB) is one of the most mature flow batteries. A flow battery stores energy in two soluble redox couples contained in external liquid electrolyte tanks. These electrolytes can be pumped from the tanks to the cell stack, which consists of two electrolyte flow compartments separated by ion selective membranes [66]. The operation is based on reduction–oxidation reactions of the electrolyte solutions. VRB stores energy by employing vanadium redox couples (V^{2+}/V^{3+} in the negative and V^{4+}/V^{5+} in the positive half-cells) in two electrolyte tanks. VRBs exploit the vanadium in these four oxidation states, which makes the flow battery have only one active element in both anolyte and catholyte. Other two types of commercially available flow batteries are the zinc bromine battery (ZBB) and the polysulfide bromide battery (PSB) [63].

VRBs have quick responses (faster than 0.001 second) and can operate for 10 000–16 000+ cycles with 100% DoD [42]. Their efficiency is relatively high, about 60–80%. The main advantage of the VRB is that it can offer almost unlimited capacity simply by using larger and larger storage tanks. The drawbacks of VRB are apparent because of its low specific energy and energy density [66–68]. VRB was pioneered at the University of New South Wales (UNSW), Australia, in the early 1980s. Nowadays, VRBs have been used for a variety of applications, such as load leveling, power quality control applications, facilitating renewable energy integration, etc.

1.2.1.9 Hydrogen Storage and Fuel Cell

Hydrogen ES systems use two separate processes for storing energy and producing electricity. The use of a water electrolysis unit is a common way to produce hydrogen, which can be stored in high-pressure containers and/or transmitted by pipelines for later use. When using the stored hydrogen for electricity generation, fuel cells are adopted [59]. Fuel cells generate electricity and heat via electrochemical reaction, which is actually the reversed electrolysis reaction. The reaction between oxygen and hydrogen to generate electricity is different for various types of fuel cell.

In general, the electricity generation by using fuel cells is quieter, produces less pollution, and is more efficient than the fossil fuel combustion approaches. Other features include easy scaling (potential from 1 kW to hundreds of MW) and compact design. Fuel cell systems combined with hydrogen production and storage can provide stationary or distributed power (primary electrical power, heating/cooling, or backup power), and transportation power (fuel cell vehicles). Notwithstanding these advantages, hydrogen fuel cells are generally somewhat at a cost disadvantage at the present time and also suffer from a relatively low round-trip efficiency [69–71]. Hydrogen may become competitive for seasonal storage of REG, which requires large energy capacity and a very low self-discharge [26]. The integration of hydrogen systems in connection with wind power generation can facilitate a high penetration of wind energy, and the hydrogen produced by surplus wind energy can be used in the transport sector [71].

1.2.1.10 Superconducting Magnetic Energy Storage

A superconducting magnetic energy storage (SMES) device is a direct current device that stores energy in the magnetic field. It is achieved by inducing direct current into a coil made of superconducting cables of nearly zero resistance [72]. A typical SMES system is composed of three main components, which include: a superconducting coil unit, a power conditioning subsystem, and a refrigeration and vacuum subsystem [59]. One advantage of SMES is its great instantaneous efficiency, near 95% for a charge–discharge cycle. The fast response time (under 100 ms) of SMES makes it ideal for regulating network stability [72]. In contrast to rechargeable batteries, SMES systems are capable of discharging near to the totality of the stored energy with little degradation after thousands of full cycles. The drawbacks are that they have high capital cost (up to 10 000\$ kWh^1, 7200\$ kW^{-1}) and a high daily self-discharge (10–15%) [59, 73].

Back in 1983, a 30 MJ (8.4 kWh) SMES unit with a 10 MW converter was installed at Bonneville Power Administration (BPA) substation in Tacoma, Washington, USA. It was designed for power system damping control [74, 75]. In reference [76], the development of SMES systems for demand-side power quality control and power system applications in Japan is introduced.

1.2.1.11 Supercapacitor

The classical capacitor comprises two parallel metal plates with an air gap between them. When a voltage is applied across the plates, positive charges are collected on one of the plates and the negative charges are accumulated on the other one. ES in supercapacitors is done in the form of an electric field between two electrodes. This is the same principle as for capacitors except that the insulating material is replaced by an electrolyte ionic conductor in which ion movement is made along a conducting electrode with a very large specific surface. Thus, supercapacitors have both the characteristics of capacitors and electrochemical batteries, except that there is no chemical reaction [72, 77].

The most important features of supercapacitors are their long cycling times, more than 100 000 cycles, and high cycle efficiency, ~84–97%. However, the daily self-discharge rate of supercapacitors is high, ~5–40%, and the capital cost is also high at present. Thus, supercapacitors are well suited for short-term storage applications but not for large-scale and long-term energy storage [52, 63, 72, 78].

1.2.1.12 Thermal Energy Storage

Thermal energy storage (TES) can store a change in internal energy of a material as sensible heat or latent heat. In sensible heat storage (SHS), thermal energy is stored by raising the temperature of a solid or liquid. Latent heat storage (LHS) is based on the heat absorption or release when a storage material undergoes a phase change from solid to liquid or liquid to gas or vice versa [79]. Based on the range of operating temperatures, TES can also be classified into two groups: low-temperature TES and high- temperature TES. The TES system can store large quantities of energy without any major hazards and its daily self-discharge loss is small (~0.05–1%). However, the cycle efficiency of TES systems is normally low (~30–60%) [59].

An LHS system using phase change materials has the advantages of high energy storage density and the isothermal nature of the storage process. The LHS that is suitable for solar heating and air conditioning has received considerable attention due to its advantages of storing a large amount of energy as a phase transition at a constant temperature [79, 80]. A concentrating solar power (CSP) system converts sunlight into a heat source that can be used to drive a conventional power plant. TES improves the flexibility of a CSP plant. Commercial deployment of CSP systems has been achieved in recent years with the two-tank sensible storage system using molten salt as the storage medium [81]. By March 2015, the CSP market had a total capacity of 5840 MW worldwide, among which 4800 MW was operational and 1040 MW was under construction.

1.2.2 Technical and Economic Analyses of Different Energy Storage Technologies

ES technology has many important technical and economic indicators/parameters, such as technology maturity, energy/power scale and density, cycle efficiency, life cycle, response time, self-discharge rate, construction cost, and life-cycle cost. Different applications have different requirements for various parameters of ES. For example, power quality control requires high power, fast response, and high life cycles, but does not require large capacity and low self-discharge rates. The ESS for grid power peak shaving and valley filling should have large power and energy capacities and high cycle efficiency, but the response time is not critical. Since there is currently no ES technology that can meet all the requirements, it is necessary to select an appropriate ES technology according to the application and the characteristics of various ES technologies. This section compares and analyzes various technical and economic indices of different ES technologies.

1.2.2.1 Power Capacity and Discharge Duration

For applications such as peak shaving and energy management, the power capacity and discharge duration of an ESS are crucial, and an ESS with small power and energy capacities cannot undertake such applications. Currently, only two technologies, PHS and CAES, can reach an installed capacity of GW and a discharging duration up to hours. Among other ES technologies, flywheel and battery ES can reach a power capacity at the MW level and discharge for several minutes to several hours. Supercapacitors and SMES have high power (up to MWs), but the discharge duration at rated power is very short (a few seconds to a few minutes).

1.2.2.2 Response Time

Most ES technologies can respond instantaneously (within 100 ms). Even with conventional generators (hydro turbine, gas turbine), PHS, and CAES have response speeds faster than that of a coal-fired unit. The rapid response is an important feature for an ES technology, making it suitable for applications such as power quality improvement, frequency regulation, stability enhancement, and intermittent renewable power smoothing.

1.2.2.3 Cycle Efficiency and Daily Standby Power Losses

The cycle efficiencies and daily standby power losses of various ES technologies are shown in Figure 1.5. The cycle efficiency of most ES technologies is above 70%. Among various ES technologies, supercapacitors, flywheels, SMES, and lithium-ion batteries have the highest cycle efficiency of over 90%. The self-discharge loss

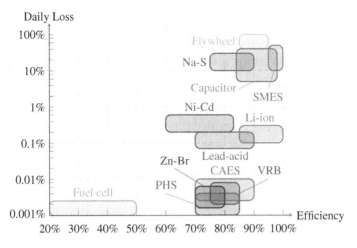

Figure 1.5 Efficiencies and daily losses of typical energy storage technologies.

of ES technology is generally small, with the exception of FES (air and mechanical friction), the sodium–sulfur battery (maintaining a high reaction temperature), the supercapacitor (leakage current), and SMES (cooling device).

1.2.2.4 Lifetime and Cycle Times

The lifetime and cycle times of various ES technologies are shown in Figure 1.6. Large-scale ESSs, i.e. PHS and CAES, have a long life, while battery ESSs have a relatively short life. In terms of cycle times, supercapacitors, flywheel, and SMES have excellent performance, up to 100 000 cycles. PHS and CAES can reach tens of

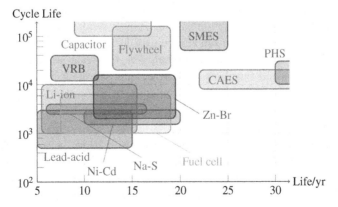

Figure 1.6 Calendar and cycle lives of typical energy storage technologies.

thousands of cycles. The cycle numbers of battery ES technologies are generally not high, among which nickel–cadmium batteries have a memory effect and their cycle times are relatively low. Lead–acid batteries also have a low cycle number. Flow batteries (such as VRBs) can cycle more than 10 000 times.

In addition, DoD has a profound impact on the number of cycles of batteries, such as the lithium-ion battery. Deep charge and discharge have a prominent negative impact on the cycle times of lithium-ion batteries, while the effect of shallow charge and discharge is trivial. For VRBs, the DoD has no significant effect on the number of cycles [82, 83]. Thus, the application characteristics should be carefully analyzed before selecting ES technologies, and a trade-off between the initial investment and life-cycle costs should also be made [84]. In reference [70], the technical parameters of typical electrical ES technologies are summarized. Table 1.2 shows the electrical ES technologies discussed in Section 1.2.1.

1.2.2.5 Construction and Operation Costs

The total capital costs of grid-scale ESSs are also given in reference [70] and part of them are listed in Table 1.3. It should be noted that costs are changing very fast for some types of ESSs, including but not limited to the lithium-ion battery, flywheel, and hydrogen storage. The per unit energy costs are relatively low for PHS and CAES due to their large scale and high cycle times. Currently, the investment costs of SMES and supercapacitors are relatively high, but their operation costs are relatively low [31]. The demand for lithium-ion batteries has increased significantly in recent years and its cost has dropped rapidly. It is pointed out in reference [85] that the price of lithium-ion batteries in 2016 was 273\$ kWh^{-1}, a drop of about 73% since 2010. The estimated cost of producing lithium-ion cells is now as low as 145\$ kWh^{-1}, which was given in reference [86], published in 2017.

1.2.3 Applications of Energy Storage in Power Systems

An ESS can be installed in different parts of a power system and undertake different roles. The same ES system can also be used for a variety of purposes through advanced power and energy management. The following sections discuss the major applications and functions of ESSs according to the locations where they are installed, i.e. generation side, transmission network, distribution network, and demands side.

1.2.3.1 Application of Large-Scale Energy Storage on Power Generation Side

1) **Independent ES to improve flexibility of power supply**

Large-scale ES has a long history as an independent power station on the power generation side. The main types are PHS power plants and CAES power stations. The main functions that they can play in the power system include peak

Table 1.2 Technical parameters of electrical energy storage system [70].

Energy storage technology	Power range (MW)	Discharge time (ms–h)	Overall efficiency	Power density (W/kg)	Energy density (Wh/kg)	Storage durability	Self-discharge (per day)	Lifetime (year)	Life cycles (cycles)
PHS	10–5000	1–24 h	0.70–0.82		0.5–1.5	h–months	Negligible	50–60	20 000–50 000
CAES (underground)	5–400	1–24 h	0.7–0.89		30–60	h–months	Small	20–40	>3000
CAES (aboveground)	3–15	2–4 h	0.70–0.90			h–days	Small	20–40	>3000
Flywheel	Up to 0.25	ms–15 m	0.93–0.95	1000	5–100	s–min	100%	15–20	20 000–100 000
Lead–acid	Up to 20	s–h	0.70–0.90	75–300	30–50	min–days	0.1–0.3%	5–15	2000–4500
Na–S	0.05–8	s–h	0.75–0.90	150–230	150–250	s–h	20%	10–15	2500–4500
Ni–Cd	Up to 40	s–h	0.60–0.73	50–1000	15–300	min–days	0.2–0.6%	10–20	2000–2500
Li–ion	Up to 0.01	m–h	0.85–0.95	50–2000	150–350	min–days	0.1–0.3%	5–15	1500–4500
VRB	0.03–3	s–10 h	0.65–0.85	166	10–35	h–months	Small	5–10	10 000–13 000
SMES	0.1–10	ms–8 s	0.95–0.98	500–2000	0.5–5	min–h	10–15%	15–20	>100 000
Capacitors	Up to 0.05	ms–60 m	0.60–0.65	100 000	0.05–5	s–h	40%	5–8	50 000
SCES[a]	Up to 0.3	ms–60 m	0.85–0.95	800–23 500	2.5–50	s–h	20–40%	10–20	>100 000
Hydrogen (fuel cell)	0.3–50	s–24 h	0.33–0.42	500	100–10 000	h–months	Negligible	15–20	20 000

[a] SCES, supercapacitor energy storage.

Table 1.3 Total capital costs of grid-scale energy storage systems [70].

Energy storage technology	Configuration	Total capital cost of per unit power rating € kW^{-1}			Total capital cost of per unit energy capacity € kWh^{-1}		
		Min	Average	Max	Min	Average	Max
PHS	Conventional	1030	1406	1675	96	137	181
CAES	Aboveground	774	893	914	48	92	106
	Underground	1286	1315	1388	210	263	278
Flywheel	High-speed	590	867	1446	1850	4791	25 049
Lead–acid	Advanced	1388	2140	3254	346	437	721
Na–S	–	1863	2254	2361	328	343	398
Ni–Cd	–	2279	3376	4182	596	699	808
Li-ion	–	2109	2512	2746	459	546	560
VRB	–	1277	1360	1649	257	307	433
Supercapacitors	Double-layer	214	229	247	691	765	856
SMES	–	212	218	568	5310	6090	6870
Hydrogen	Fuel cell (FC)	2395	3243	4674	399	540	779

shaving and valley filling, participation in frequency regulation and voltage control, provision of reserve, etc. They are reliable, economical, long life cycle, large capacity, and mature ES technologies.

2) **Installing ES alongside a renewable power plant**

ESSs, such as batteries, supercapacitors, and flywheels, have a fast-response performance. They can be installed alongside renewable power plants, which can smooth the fluctuating renewable power output [87, 88], improve their controllability, and enhance their competitiveness in electricity markets [26, 89, 90]. In the northern part of China, due to insufficient capacity of both transmission lines and flexible generators, a large amount of renewable energy was spilled in recent years [91]. If ESSs were installed within or near to renewable power plants, the curtailed renewable energy could be effectively reduced through the optimal power and energy control of the storage systems. Table 1.4 summarizes the typical applications and requirements for ES to support REG.

1.2.3.2 Install Large-Scale Energy Storage Systems for Power Transmission

A transmission network often covers a large area with high voltage level(s) and large transmission capacities, so only large-scale ESSs can be installed to the transmission network. They can play a major role in the following two aspects:

Table 1.4 Applications of energy storage for renewable energy generation.

Functions of ES	Requirements for ES	Suitable ES types	Explanation
Smoothing power	Power: ~100 kW-MWs Duration: minutes to hours Response time: milliseconds	Battery, supercapacitor, flywheel	By tracking and smoothing the power changes of REG, meets the operational requirements and improving friendliness of grid connection for REG.
Following scheduled power	Power: ~100 kW-MWs Duration: hours Response time: minutes	Battery, thermal energy storage, compressed air energy storage, hydrogen storage	By deploying large-scale ESS for REG station, their total output can follow the scheduled profile based on REG power forecast and reasonable charging/discharging dispatch of ESS.
Reducing REG curtailment	Power: MWs Duration: hours Response time: minutes	Battery, thermal energy storage, hydrogen storage	When the REG output cannot be fully accepted by the power system, the ESS stores the surplus electricity and the stored energy can be released back to the power grid when allowed.
Price arbitrage	Power: MWs Duration: hours Response time: minutes	Battery, thermal storage, compressed air energy storage, hydrogen storage	Under the electricity market environment, the electricity price of power generation changes with time. The ESS charges during the low-price time period and discharges during the high-price time period, which profits through price arbitrage.
Providing ancillary services	Power: MWs Duration: 10 minutes to hours Response time: seconds	Battery, flywheel, compressed air energy storage, etc.	With fast response capability, the ESS can provide the ancillary services such as primary and secondary frequency regulation, reserve, and ramping for the whole power system when needed.

(Continued)

Table 1.4 (Continued)

Functions of ES	Requirements for ES	Suitable ES types	Explanation
Enhancing stability	Power: ~100 kW-MWs Duration: minutes Response time: milliseconds	Battery, supercapacitor, flywheel, superconducting magnetic energy storage	With well-tuned frequency/voltage control strategies, the ESS can provide large active/ reactive power support quickly in the event of power grid contingency.

(a) As an alternative to transmission expansion or to upgrade investment (deferring or reducing the investment on a transmission network) [92], improve the transmission capacity of key transmission corridors [93] or improve power system stability [94].

(b) System level applications, such as providing different types of ancillary service including frequency regulation, reserve, reactive power/voltage support, black start, etc. [95].

1) **As an alternative to transmission expansion or upgrade investment**
   The increase of load power and integration of power plants (especially the integration of large-scale renewable energy generation) require new transmission lines and substations to increase the transmission capacity of the power system. However, due to the constraints of land use and environment protection, it is becoming more and more difficult to build new lines and substations. Even after getting permission, it usually takes years to build these new facilities. Thus, it cannot meet the needs of rapid development of REG and load growth. Large-scale ESSs can be used as new means to be installed in transmission networks to enhance the transmission capacity and reduce investment in lines and substations.

   It should be noted that PHS stations and large-scale CAES stations are typically restricted by geological conditions. Battery storage systems are becoming an option in recent years. At the planning stage of a transmission network, a comprehensive cost–benefit analysis needs to be properly carried out from the view of the entire power system by taking into account both transmission network expansion and ESS deployment.

2) **Providing ancillary services**
   As mentioned before, a large-scale ESS can provide frequency regulation, reserve, reactive power/voltage support, and black start services for a bulk

power system. Compared with traditional generating units, emerging ES technologies such as battery ES and flywheel have the advantages of prominent ramping ability and a fast response speed, which can quickly compensate for the changes of system load and intermittent renewable power output [96]. Although the energy capacity of an ESS may be relatively small, this shortcoming is not a problem because it tracks signals with fast fluctuations, small amplitudes, and average values close to zero in frequency regulation applications. These characteristics give the ESSs natural performance advantages in frequency regulation and competitiveness in the frequency regulation service market. The participation of ESS in providing frequency regulation helps to alleviate the extra requirement of regulation capacity brought about by the large-scale integration of renewable energy generation [97].

The benefits that can be obtained from investing in the ESS include price arbitrage, providing ancillary services, participation in the capacity market and reducing/deferring network investment, etc. The multiple functions an ESS can play should be coordinated and optimally controlled.

1.2.3.3 Install Energy Storage Systems in the Distribution Network

The voltage levels of distribution networks are relatively low. The required power and energy capacities of an ESS are lower than those for transmission networks. The roles that ESSs can play for distribution networks include improving security [98], economy [99], reliability [100], the ability to integrate more DG [101], etc. In addition, ESSs connected to distribution networks can also provide ancillary services such as frequency regulation and reserve for the whole power system.

1) **Reduce or defer the distribution network upgrade investment**
   As the load increases, when the capacity of the distribution transformer or line is insufficient, the traditional approach is to increase the corresponding capacity by replacing the existing equipment with larger capacity ones or add new ones. However, the investment in transformers and lines may be large, the construction period is often long, or spaces for new facilities are not available. Adding ESSs to distribution networks can reduce the peak load of transformers and distribution lines by discharging power during peak load periods, thus achieving the goal of reducing or deferring the upgrade investment of distribution networks.

2) **Guarantee secure and economic operation of the distribution system**
   With an ESS installed in a distribution network, its four-quadrant control of active and reactive power can be performed to change the power flow in the distribution network, thereby improving the voltage quality and reducing the active power losses. In addition, price arbitrage can be achieved by charging during low electricity price periods and discharging at high price periods.

3) **Improve the reliability of the power supply**
Distribution network failures has primary responsibility for power outages of customers. When a fault within the distribution network causes some users to be isolated and disconnected from the network, the local ESS can supply power to these users. This will greatly reduce the power outage time and even realize uninterrupted power supply for the users, thereby improving the power supply reliability.

4) **Boost the integration of distributed generation**
In recent years, with the development of DG technology and the support of government policies, the installed capacity of DG has been continuously increased. In particular, the rapid development of distributed PV power generation brings new problems and challenges to the secure operation of distribution networks. In some areas with abundant solar radiation and wind resources, the distribution network may be weak, requiring new technologies and control means to ensure secure and reliable integration of REG. ESSs, which can be easily modularized, installed, and relocated, are effective means to boost the integration of DG. In addition, the distributed generators and ESSs can be optimally sized and connected to the distribution network simultaneously, which will improve the controllability of distributed REG.

5) **Providing ancillary services for the bulk power system**
With the permission of relevant policies and electricity market rules, the ESSs installed in the distribution networks can provide ancillary services such as frequency adjustment and reserve for the bulk power system. Although the capacity of a single ESS may be small, the aggregation of multiple distributed ESSs can result in considerable fast frequency regulation and reserve capacity.

With the development of electric vehicles, the on-board battery of electric vehicles has been regarded as a potential ES resource. In addition to a large amount of research conducted by the academic community, some pilot and commercial projects have been launched to show the feasibility of vehicle-to-grid (V2G) applications for power balance and frequency regulation. In addition, a large number of retired batteries from electric vehicles can be further utilized to provide considerable ES capacity for power systems [102].

1.2.3.4 Demand-Side Energy Storage

Nowadays, important electricity users or loads are usually equipped with a UPS. In the case of a power outage, the UPS can independently operate for a period of time, which can greatly reduce the power interruption cost. With the advancement of ES technologies, ES will gain more and more applications at the demand side [103, 104], including increasing power supply reliability, reducing electricity costs (through price arbitrage), and facilitating distributed REG, participating demand response, improving power quality, etc.

1) **Reduce cost and improve power supply reliability for industrial and commercial users**

The types of ES that can be deployed by industrial and commercial customers include chemical batteries, TES, flywheel, etc. Price arbitrage and participation in demand side responses are the main types of applications for these ESSs. For price arbitrage, time of use or dynamic electricity prices are required. Part or all of the electricity consumption can be transferred by ES from peak to valley price time periods. For the demand side response, the charging and discharging of the stored energy is controlled to meet the dispatch schedule from the power system operator [105].

2) **Improve controllability of demand-side DG**

For industrial and commercial users, by installing PV system on the roofs of their factories, office buildings, or parking lots, their costs on purchasing electricity can be reduced and they may even sell electricity to the grid or other users. By investing in ESSs, it is possible to stabilize the volatility of PV power output, improve power quality, and implement price arbitrage [106]. Due to the requirements of flexible power control, the battery storage systems are typically the first choice.

3) **Improve reliability and coordinate with DG for residential users**

PV panels generate electricity during the daytime, while residential power load generally reaches its peak at night. By installing an ESS and rooftop PV system at home, the electricity generated by the PV system can be saved by the ES and released for home usage during the night, which can even achieve self-sufficiency in electricity for a home [107]. In the market, there are more and more products for household ES (such as Tesla's Powerwall). In the case of a distribution network contingency, the household ES can continue to supply power, thereby effectively reducing the impact of power outage and improving the reliability of the power supply.

It is worth mentioning that electric vehicles can also play a role-like household ES through vehicle-to-home (V2H) technology [108]. In the case of power failure of the distribution network, the power battery of an electric vehicle can supply power to the household electrical equipment through an energy conversion device.

1.3 Chapter Structure

This book mainly focuses on the optimal operation and planning of ES that may be installed in different parts of a power system. The content of all the following chapters are briefly introduced as follows:

Chapter 2. Chapter 2 mainly introduces the working principles and steady-state models of three types of ESSs, i.e. PHS, CAES, and battery ES.

Chapter 3. This chapter discusses the optimal formulations for the coordination of an ESS with renewable energy generation. The day-ahead dispatch model is built considering the randomness of renewable power output. In addition, the bidding strategy of the union of ES and renewable energy generation is also discussed.

Chapter 4. The optimal operating and bidding strategies are further studied in this chapter. Chapter 4 consists of two parts. Based on the linear decision rules, the first part describes a stochastic optimization strategy of the ES and renewable energy generation union for its participation in the power market. The second part provides a rolling optimization scheme, which includes the day-ahead and intra-day bidding strategies and the real-time control strategy for the union.

Chapter 5. This chapter discusses the modeling and optimization of large-scale ESSs participating in UC. The deterministic, stochastic, and robust optimization models are respectively formulated for UC with ESSs.

Chapter 6. In this chapter, the optimal power flow (OPF) formulation is extended as a multiple time-interval optimization problem considering the optimal charging and discharging of large-scale ESSs. The conventional nonlinear programming and semidefinite programming (SDP) models are compared.

Chapter 7. This chapter mainly discusses the control strategy of ES participating in the automatic generation control (AGC) or secondary frequency regulation of large-scale power systems. The contributions from fast response ES to the performance of AGC are analyzed.

Chapter 8. This chapter deals with the problem of transmission expansion planning considering investment in large-scale ESSs. The key points are the mathematical model and solution method for the planning problem with ES.

Chapter 9. This last chapter aims at the optimal siting and sizing problem for installing ES in distribution networks. Considering the output uncertainty of DG, the planning model and solution method for deploying distributed ES are provided. The joint planning of the distribution network and ES is further discussed.

The organization of the chapters is illustrated in Figure 1.7.

1.4 Notes to Readers

1.4.1 Topics Not Included in This Book

This book introduces the application of large-scale ES for the optimal operation and planning of power systems. Some topics that are related to applications of ES in power systems but not covered in this book are as follows:

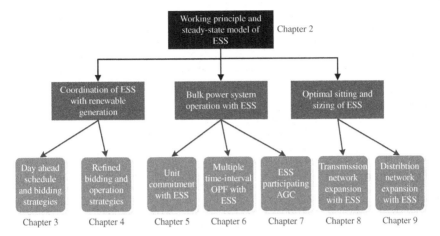

Figure 1.7 Structure illustration of this book from Chapters 2 to 9.

1) Dynamic models of ESSs. Chapter 2 of this book will present the steady-state models of several ES technologies without introducing the dynamic models of them. A pioneer work is presented in reference [109] on the dynamic model of the battery ESS. Reference [110] compares the dynamic models of battery ES for grid frequency control. Reference [111] proposes a generalized model of ESSs for power system transient stability analysis.

2) Hybrid ESS. The optimal operation and planning of the ESS discussed in this book does not consider the coordination of multiple types of ESSs. Hybrid ESSs generally consist of an ES (such as flywheel, supercapacitor, or SMES) with a high response speed and high power density, and another ES with a high energy capacity (such as electrochemical ES). By optimal coordination between different ESSs, a better control performance can be achieved for both power profile smoothing and energy shifting than that of a single-type ES [112, 113].

3) Improving transient stability by ES. An ESS with a fast power response capability can be used to improve the transient stability of power systems. For large-scale power systems, a large power capacity of the ESS is generally required. The key points for research include the model building and control strategy of the ESSs [114–116].

4) ESS within a microgrid. For a relatively small-scale microgrid (or independent grid), if the uncertain wind/solar power generation accounts for a relatively high proportion, it is necessary to install ES for the power balance and stability control [117, 118]. The optimal scheduling and control strategy for ES with renewable energy generation discussed in this book can be used for microgrid operation with proper adaption, but the relevant content in this book is not specific for the microgrid.

1.4.2 Required Basic Knowledge

The contents of the subsequent chapters of this book will be unfolded without providing the related basic knowledge of power system operation and planning. The readers may refer to references [119] and [120] for knowledge of renewable energy generation and the electricity market. The optimal operation and scheduling of power systems can be found in reference [121], and reference [122] describes the models and solution methods for power system planning.

References

1 GWEC (April 2019). Global Wind Report 2018. www.gwec.net.

2 IEA Photovoltaic Power Systems Programme (2019). Snapshot of Global PV Markets. Report IEA PVPS T1–35: 2019. www.iea-pvps.org.

3 IEA Photovoltaic Power Systems Programme (2018). Trends 2018 in Photovoltaic Applications. Report IEA PVPS T1–34: 2018. www.iea-pvps.org.

4 Holttinen, H., Meibom, P., Orths, A. et al. (2011). Impacts of large amounts of wind power on design and operation of power systems, results of IEA collaboration. *Wind Energy* 14 (2): 179–192.

5 Denholm, P. and Hand, M. (2011). Grid flexibility and storage required to achieve very high penetration of variable renewable electricity. *Energy Policy* 39 (3): 1817–1830.

6 Riesz, J. and Milligan, M. (2015). Designing electricity markets for a high penetration of variable renewables. *Wiley Interdisciplinary Reviews: Energy and Environment* 4 (3): 279–289.

7 Wang, Y., Silva, V., and Lopez-Botet-Zulueta, M. (2016). Impact of high penetration of variable renewable generation on frequency dynamics in the continental Europe interconnected system. *IET Renewable Power Generation* 10 (1): 10–16.

8 Eftekharnejad, S., Vittal, V., Heydt, G.T. et al. (2013). Small signal stability assessment of power systems with increased penetration of photovoltaic generation: A case study. *IEEE Transactions on Sustainable Energy* 4 (4): 960–967.

9 Flynn, D., Rather, Z., Ardal, A. et al. (2017). Technical impacts of high penetration levels of wind power on power system stability. *Wiley Interdisciplinary Reviews: Energy and Environment* 6 (2): 1–19.

10 Palmintier, B.S. and Webster, M.D. (2016). Impact of operational flexibility on electricity generation planning with renewable and carbon targets. *IEEE Transactions on Sustainable Energy* 7 (2): 672–684.

11 Pina, A., Silva, C.A., and Ferrão, P. (2013). High-resolution modeling framework for planning electricity systems with high penetration of renewables. *Applied Energy* 112: 215–223.

12 Romero, R., Monticelli, A., Garcia, A. et al. (2002). Test systems and mathematical models for transmission network expansion planning. *IEE Proceedings-Generation, Transmission and Distribution* 149 (1): 27–36.

13 Zhang, H., Heydt, G.T., Vittal, V., and Quintero, J. (2013). An improved network model for transmission expansion planning considering reactive power and network losses. *IEEE Transactions on Power Systems* 28 (3): 3471–3479.

14 Buygi, M.O., Balzer, G., Shanechi, H.M., and Shahidehpour, M. (2004). Market-based transmission expansion planning. *IEEE Transactions on Power Systems* 19 (4): 2060–2067.

15 Zhao, J.H., Dong, Z.Y., Lindsay, P., and Wong, K.P. (2009). Flexible transmission expansion planning with uncertainties in an electricity market. *IEEE Transactions on Power Systems* 24 (1): 479–488.

16 Lumbreras, S. and Ramos, A. (2016). The new challenges to transmission expansion planning. Survey of recent practice and literature review. *Electric Power Systems Research* 134: 19–29.

17 Jabr, R.A. (2013). Robust transmission network expansion planning with uncertain renewable generation and loads. *IEEE Transactions on Power Systems* 28 (4): 4558–4567.

18 Orfanos, G.A., Georgilakis, P.S., and Hatziargyriou, N.D. (2013). Transmission expansion planning of systems with increasing wind power integration. *IEEE Transactions on Power Systems* 28 (2): 1355–1362.

19 Ackermann, T., Andersson, G., and Söder, L. (2001). Distributed generation: a definition. *Electric Power Systems Research* 57 (3): 195–204.

20 Guerrero, J.M., Blaabjerg, F., Zhelev, T. et al. (2010). Distributed generation: Toward a new energy paradigm. *IEEE Industrial Electronics Magazine* 4 (1): 52–64.

21 Dugan, R.C., McDermott, T.E., and Ball, G.J. (2001). Planning for distributed generation. *IEEE Industry Applications Magazine* 7 (2): 80–88.

22 Mahmoud Pesaran, H.A., Huy, P.D., and Ramachandaramurthy, V.K. (2017). A review of the optimal allocation of distributed generation: objectives, constraints, methods, and algorithms. *Renewable and Sustainable Energy Reviews* 75: 293–312.

23 Haffner, S., Pereira, L.F.A., Pereira, L.A. et al. (2008). Multistage model for distribution expansion planning with distributed generation – Part I: Problem formulation. *IEEE Transactions on Power Delivery* 23 (2): 915–923.

24 Haffner, S., Pereira, L.F.A., Pereira, L.A. et al. (2008). Multistage model for distribution expansion planning with distributed generation – Part II: Numerical results. *IEEE Transactions on Power Delivery* 23 (2): 924–929.

25 Muñoz-Delgado, G., Contreras, J., and Arroyo, J.M. (2015). Joint expansion planning of distributed generation and distribution networks. *IEEE Transactions on Power Systems* 30 (5): 2579–2590.

26 Beaudin, M., Zareipour, H., Schellenberglabe, A. et al. (2010). Energy storage for mitigating the variability of renewable electricity sources: An updated review. *Energy for Sustainable Development* 14 (4): 302–314.

27 Taylor, J., Callaway, D.S., and Poolla, K. (May 2013). Competitive energy storage in the presence of renewables. *IEEE Transactions on Power Systems* 28 (2): 985–996.

28 Rodrigues, E.M.G., Godina, R., Santos, S.F. et al. (2014). Energy storage systems supporting increased penetration of renewables in islanded systems. *Energy* 75: 265–280.

29 Dell, R.M. and Rand, D.A.J. (2001). Energy storage – a key technology for global energy sustainability. *Journal of power sources* 100 (1–2): 2–17.

30 Bussar, C., Moos, M., Alvarez, R. et al. (2014). Optimal allocation and capacity of energy storage systems in a future European power system with 100% renewable energy generation. *Energy Procedia* 46: 40–47.

31 SANDRIA (July 2013). SANDIA Report, DOE/EPRI 2013 Electricity Storage Handbook in Collaboration with NRECA, SAND2013–5131. https://www.sandia. gov/ess-ssl/lab_pubs/doeepri-electricity-storage-handbook.

32 Rehman, S., Al-Hadhrami, L.M., and Alam, M.M. (2015). Pumped hydro energy storage system: A technological review. *Renewable and Sustainable Energy Reviews* 44: 586–598.

33 Du, P. and Lu, N. (2014). *Energy Storage for Smart Grids: Planning and Operation for Renewable and Variable Energy Resources (VERs)*. Academic Press.

34 DOE Global Energy Storage Database (February 2019). www. energystorageexchange.org.

35 International Electrotechnical Commission (IEC) (2011). Electrical energy storage: white paper. Technical Report, International Electrotechnical Commission (IEC). http://www.iec.ch/whitepaper/pdf/iecWP-energystorage-LR-en.pdf.

36 Dominion Energy (February 2019). Bath County Pumped Storage Station (EB/OL), 16 February 2019. https://www.dominionenergy.com/about-us/making-energy/renewable-generation/water/bath-county-pumped-storage-station.

37 Chen, H., Ngoc, T., Yang, W. et al. (2009). Progress in electrical energy storage system: A critical review. *Progress in Natural Science* 19 (3): 291–312.

38 Fujihara, T., Imano, H., and Oshima, K. (1998). Development of pump turbine for seawater pumped-storage power plant. *Hitachi Review* 47 (5): 199–202.

39 Pickard, W.F. (2012). The history, present state, and future prospects of underground pumped hydro for massive energy storage. *Proceedings of the IEEE* 100 (2): 473–483.

40 Yang, C.J. and Jackson, R.B. (2011). Opportunities and barriers to pumped-hydro energy storage in the United States. *Renewable and Sustainable Energy Reviews* 15: 839–844.

41 Crotogino, F., Mohmeyer, K.U., and Scharf, R. *Huntorf CAES: More than 20 Years of Successful Operation*. Spring 2001 Meeting, Orlando, FL, USA, 2001. pp. 1–7. CAES.

42 Luo, X., Wang, J., Dooner, M. et al. (2014). Overview of current development in compressed air energy storage technology. *Energy Procedia* 62: 603–611.

43 Budt, M., Wolf, D., Span, R., and Yan, J. (2016). A review on compressed air energy storage: Basic principles, past milestones and recent developments. *Applied Energy* 170: 250–268.

44 Zunft, S., Dreissigacker, V., Bieber, M., et al. Electricity storage with adiabatic compressed air energy storage: Results of the BMWi-project ADELE-ING. *International ETG Congress 2017* (28 November 2017), pp. 1–5. VDE.

45 Liu, H. and Jiang, J. (2007). Flywheel energy storage – An upswing technology for energy sustainability. *Energy and Buildings* 39 (5): 599–604.

46 Faraji, F., Majazi, A., and Al-Haddad, K. (2017). A comprehensive review of flywheel energy storage system technology. *Renewable and Sustainable Energy Reviews* 67: 477–490.

47 Díaz-González, F., Sumper, A., Gomis-Bellmunt, O. et al. (2013). Energy management of flywheel-based energy storage device for wind power smoothing. *Applied Energy* 110: 207–219.

48 Suzuki, Y., Koyanagi, A., Kobayashi, M. et al. (2005). Novel applications of the flywheel energy storage system. *Energy* 30 (11–12): 2128–2143.

49 Sun, B., Dragičević, T., Freijedo, F.D. et al. (2016). A control algorithm for electric vehicle fast charging stations equipped with flywheel energy storage systems. *IEEE Transactions on Power Electronics* 31 (9): 6674–6685.

50 Abdeltawab, H.H. and Mohamed, Y.A. (2016). Robust energy management of a hybrid wind and flywheel energy storage system considering flywheel power losses minimization and grid-code constraints. *IEEE Transactions on Industrial Electronics* 63 (7): 4242–4254.

51 Dunn, B., Kamath, H., and Tarascon, J.M. (2011). Electrical energy storage for the grid: a battery of choices. *Science* 334 (6058): 928–935.

52 Hadjipaschalis, I., Poullikkas, A., and Efthimiou, V. (2009). Overview of current and future energy storage technologies for electric power applications. *Renewable and Sustainable Energy Reviews* 13 (6–7): 1513–1522.

53 Baker, J. (2008). New technology and possible advances in energy storage. *Energy Policy* 36 (12): 4368–4373.

54 Doughty, D.H., Butler, P.C., Akhil, A.A. et al. (2010). Batteries for large-scale stationary electrical energy storage. *Electrochemical Society Interface*: 49–53.

55. Rand, D., Garche, J., Moseley, P. et al. (2004). *Valve-Regulated Lead-Acid Batteries*. Amsterdam, Netherlands: Taylor & Francis.

56 EPRI (December 2009). *Energy Storage and Distributed Generation Technology Assessment: Assessment of Lead–Acid–Carbon, Advanced Lead–Acid, and Zinc-Air–Batteries for Stationary Application*, EPRI ID 1017811. Palo Alto, CA: EPRI.

57 Wehner, J., Mohler, D., Gibson, S., et al. (2015). Technology Performance Report: Duke Energy Notrees Wind Storage Demonstration Project. Duke Energy

Renewables, Charlotte, NC, USA. DOI: 10.2172/1369566, https://www.osti.gov/servlets/purl/1369566.

58 McKeon, B.B., Furukawa, J., and Fenstermacher, S. (2014). Advanced lead–acid batteries and the development of grid-scale energy storage systems. *Proceedings of the IEEE* 102 (6): 951–963.

59 Luo, X., Wang, J., Dooner, M. et al. (2015). Overview of current development in electrical energy storage technologies and the application potential in power system operation. *Applied Energy* 137: 511–536.

60 Divya, K.C. and Østergaard, J. (2009). Battery energy storage technology for power systems – An overview. *Electric Power Systems Research* 79 (4): 511–520.

61 Poullikkas, A. (2013). A comparative overview of large-scale battery systems for electricity storage. *Renewable and Sustainable Energy Reviews* 27: 778–788.

62 Oshima, T., Kajita, M., and Okuno, A. (2004). Development of sodium–sulfur batteries. *International Journal of Applied Ceramic Technology* 1 (3): 269–276.

63 Díaz-González, F., Sumper, A., Gomis-Bellmunt, O. et al. (2012). A review of energy storage technologies for wind power applications. *Renewable and Sustainable Energy Reviews* 16 (4): 2154–2171.

64 Kawakami, N., Iijima, Y., Sakanaka, Y., et al. (2010). Development and field experiences of stabilization system using 34MW NaS batteries for a 51MW wind farm. *IEEE International Symposium on Industrial Electronics (ISIE)*, Bari, Indonesia, 2010, pp. 2371–2376.

65 Wakihara, M. (2001). Recent developments in lithium ion batteries. *Materials Science and Engineering: R: Reports* 33 (4): 109–134.

66 Leung, P., Li, X., Leon, C. et al. (2012). Progress in redox flow batteries, remaining challenges and their applications in energy storage. *RSC Advances* 2 (27): 10125–10156.

67 Barton, J.P. and Infield, D.G. (2004). Energy storage and its use with intermittent renewable energy. *IEEE Transactions on Energy Conversion* 19 (2): 441–448.

68 Sasaki, T., Kadoya, T., and Enomoto, K. (2004). Study on load frequency control using Redox flow batteries. *IEEE Transactions on Power Systems* 19 (1): 660–667.

69 Mekhilef, S., Saidur, R., and Safari, A. (2012). Comparative study of different fuel cell technologies. *Renewable and Sustainable Energy Reviews* 16 (1): 981–989.

70 Zakeri, B. and Syri, S. (2015). Electrical energy storage systems: A comparative life cycle cost analysis. *Renewable and Sustainable Energy Reviews* 42: 569–596.

71 González, A., McKeogh, E., and Gallachóir, B.O. (2004). The role of hydrogen in high wind energy penetration electricity systems: The Irish case. *Renewable Energy* 29 (4): 471–489.

72 Ibrahim, H., Ilinca, A., and Perron, J. (2008). Energy storage systems – Characteristics and comparisons. *Renewable and Sustainable Energy Reviews* 12 (5): 1221–1250.

73 Ali, M., Wu, B., and Dougal, R.A. (2010). An overview of SMES applications in power and energy systems. *IEEE Transactions on Sustainable Energy* 1 (1): 38–47.

74 Boenig, H.J. and Hauer, J.F. (1985). Commissioning tests of the Bonneville power administration 30 MJ superconducting magnetic energy storage unit. *IEEE Transactions on Power Apparatus and Systems* PAS-104 (2): 302–312.

75 Hassenzahl, W.V., Hazelton, D.W., Johnson, B.K. et al. (2004). Electric power applications of superconductivity. *Proceedings of the IEEE* 92 (10): 1655–1674.

76 Katagiri, T., Nakabayashi, H., Nijo, Y. et al. (2009). Field test result of 10MVA/20MJ SMES for load fluctuation compensation. *IEEE Transactions on Applied Superconductivity* 19 (3): 1993–1998.

77 Sharma, P. and Bhatti, T.S. (2010). A review on electrochemical double-layer capacitors. *Energy Conversion and Management* 51 (12): 2901–2912.

78 Abbey, C. and Joos, G. (2007). Supercapacitor energy storage for wind energy applications. *IEEE Transactions on Industry Applications* 43 (3): 769–776.

79 Sharma, A., Tyagi, V.V., Chen, C.R. et al. (2009). Review on thermal energy storage with phase change materials and applications. *Renewable and Sustainable Energy Reviews* 13 (2): 318–345.

80 Demirbas, M.F. (2006). Thermal energy storage and phase change materials: an overview. *Energy Sources, Part B: Economics, Planning, and Policy* 1 (1): 85–95.

81 Liu, M., Tay, N.S., Bell, S. et al. (2016). Review on concentrating solar power plants and new developments in high temperature thermal energy storage technologies. *Renewable and Sustainable Energy Reviews* 53: 1411–1432.

82 Guena, T. and Leblanc, P. (2006). How depth of discharge affects the cycle life of lithium–metal–polymer batteries. *28th International Telecommunications Energy Conference (INTELEC 06)*, Providence, RI, USA: IEEE, pp. 1–8.

83 Ibrahim, H., Ilinca, A., and Perron, J. (2007). Comparison and analysis of different energy storage techniques based on their performance index. *2007 IEEE Canada Electrical Power Conference*, Montreal, QC, Canada: IEEE, pp. 393–398.

84 Chawla, M., Naik, R., Burra, R., et al. (2010). Utility energy storage life degradation estimation method. *IEEE Conference on Innovative Technologies for an Efficient and Reliable Electricity Supply (CITRES)*, Waltham, MA, USA: IEEE, pp. 302–308.

85 Curry, C. (2007). Lithium-ion battery costs and market. *Bloomberg New Energy Finance*.

86 Blomgren, G.E. (2017). The development and future of lithium ion batteries. *Journal of The Electrochemical Society* 164 (1): A5019–A5025.

87 Evans, A., Strezov, V., and Evans, T.J. (2012). Assessment of utility energy storage options for increased renewable energy penetration. *Renewable and Sustainable Energy Reviews* 16 (6): 4141–4147.

88 Suberu, M.Y., Mustafa, M.W., and Bashir, N. (2014). Energy storage systems for renewable energy power sector integration and mitigation of intermittency. *Renewable and Sustainable Energy Reviews* 35: 499–514.

89 Weitemeyer, S., Kleinhans, D., Vogt, T., and Agert, C. (2015). Integration of renewable energy sources in future power systems: the role of storage. *Renewable Energy* 75: 14–20.

90 Parastegari, M., Hooshmand, R.A., Khodabakhshian, A., and Zare, A.H. (2015). Joint operation of wind farm, photovoltaic, pump-storage and energy storage devices in energy and reserve markets. *International Journal of Electrical Power & Energy Systems* 64: 275–284.

91 Zhang, N., Lu, X., McElroy, M.B. et al. (2016). Reducing curtailment of wind electricity in China by employing electric boilers for heat and pumped hydro for energy storage. *Applied Energy* 184: 987–994.

92 Qiu, T., Xu, B., Wang, Y. et al. (2017). Stochastic multistage coplanning of transmission expansion and energy storage. *IEEE Transactions on Power Systems* 32 (1): 643–651.

93 Del Rosso, A.D. and Eckroad, S.W. (2014). Energy storage for relief of transmission congestion. *IEEE Transactions on Smart Grid* 5 (2): 1138–1146.

94 Fang, J., Yao, W., Chen, Z. et al. (2014). Design of anti-windup compensator for energy storage-based damping controller to enhance power system stability. *IEEE Transactions on Power Systems* 29 (3): 1175–1185.

95 Kim, W.W., Shin, J.S., and Kim, J.O. (November 2017). Operation strategy of multi-energy storage system for ancillary services. *IEEE Transactions on Power Systems* 32 (6): 4409–4417.

96 Cheng, B. and Powell, W.B. (2018). Co-optimizing battery storage for the frequency regulation and energy arbitrage using multi-scale dynamic programming. *IEEE Transactions on Smart Grid* 9 (3): 1997–2005.

97 He, G., Chen, Q., Kang, C. et al. (2016). Optimal bidding strategy of battery storage in power markets considering performance-based regulation and battery cycle life. *IEEE Transactions on Smart Grid* 7 (5): 2359–2367.

98 Nick, M., Cherkaoui, R., and Paolone, M. (2014). Optimal allocation of dispersed energy storage systems in active distribution networks for energy balance and grid support. *IEEE Transactions on Power Systems* 29 (5): 2300–2310.

99 Deeba, S.R., Sharma, R., Saha, T.K. et al. (2016). Evaluation of technical and financial benefits of battery-based energy storage systems in distribution networks. *IET Renewable Power Generation* 10 (8): 1149–1160.

100 Saboori, H., Hemmati, R., and Jirdehi, M.A. (2015). Reliability improvement in radial electrical distribution network by optimal planning of energy storage systems. *Energy* 93: 2299–2312.

101 Hill, C.A., Such, M.C., Chen, D. et al. (2012). Battery energy storage for enabling integration of distributed solar power generation. *IEEE Transactions on Smart Grid* 3 (2): 850–857.

102 Liu, J., Hu, Z., Banister, D. et al. (2018). The future of energy storage shaped by electric vehicles: A perspective from China. *Energy* 154: 249–257.

103 Wang, Z., Gu, C., Li, F. et al. (2013). Active demand response using shared energy storage for household energy management. *IEEE Transactions on Smart Grid* 4 (4): 1888–1897.

104 Joo, I.Y. and Choi, D.H. (2017). Distributed optimization framework for energy management of multiple smart homes with distributed energy resources. *IEEE Access* 5: 15551–15560.

105 Wang, Y., Wang, B., Chu, C.C. et al. (2016). Energy management for a commercial building microgrid with stationary and mobile battery storage. *Energy and Buildings* 116: 141–150.

106 Parra, D., Gillott, M., Norman, S.A. et al. (2015). Optimum community energy storage system for PV energy time-shift. *Applied Energy* 137: 576–587.

107 Erdinc, O., Paterakis, N.G., Pappi, I.N. et al. (2015). A new perspective for sizing of distributed generation and energy storage for smart households under demand response. *Applied Energy* 143: 26–37.

108 Liu, C., Chau, K.T., Wu, D. et al. (2013). Opportunities and challenges of vehicle-to-home, vehicle-to-vehicle, and vehicle-to-grid technologies. *Proceedings of the IEEE* 101 (11): 2409–2427.

109 Lu, C.F., Liu, C.C., and Wu, C.J. (1995). Dynamic modelling of battery energy storage system and application to power system stability. *IEE Proceedings – Generation, Transmission and Distribution* 142 (4): 429–435.

110 Adrees, A., Andami, H., and Milanović, J.V. (2016). Comparison of dynamic models of battery energy storage for frequency regulation in power system. *2016 18th Mediterranean Electrotechnical Conference (MELECON)*. IEEE, pp. 1–6.

111 Ortega, A. and Milano, F. (2016). Generalized model of VSC-based energy storage systems for transient stability analysis. *IEEE Transactions on Power Systems* 31 (5): 3369–3380.

112 Ise, T., Kita, M., and Taguchi, A. (2005). A hybrid energy storage with a SMES and secondary battery. *IEEE Transactions on Applied Superconductivity* 15 (2): 1915–1918.

113 Choi, M.E., Kim, S.W., and Seo, S.W. (2012). Energy management optimization in a battery/supercapacitor hybrid energy storage system. *IEEE Transactions on Smart Grid* 3 (1): 463–472.

114 Chandra, S., Gayme, D.F., and Chakrabortty, A. (2014). Coordinating wind farms and battery management systems for inter-area oscillation damping: A frequency-domain approach. *IEEE Transactions on Power Systems* 29 (3): 1454–1462.

115 Ortega, A. and Milano, F. (2018). Stochastic transient stability analysis of transmission systems with inclusion of energy storage devices. *IEEE Transactions on Power Systems* 33 (1): 1077–1079.

116 Gil-González, W., Montoya, O.D., and Garces, A. (2018). Control of a SMES for mitigating subsynchronous oscillations in power systems: A PBC-PI approach. *Journal of Energy Storage* 20: 163–172.

117 Chen, S.X., Gooi, H.B., and Wang, M.Q. (2012). Sizing of energy storage for microgrids. *IEEE Transactions on Smart Grid* 3 (1): 142–151.

118 Fathima, A.H. and Palanisamy, K. (2015). Optimization in microgrids with hybrid energy systems – A review. *Renewable and Sustainable Energy Reviews* 45: 431–446.

119 Lin, J. and Magnago, F.H. (2017). *Electricity Markets: Theories and Applications*. Wiley.

120 Morales, J.M., Conejo, A.J., Madsen, H. et al. (2013). *Integrating Renewables in Electricity Markets: Operational Problems*. Springer Science and Business Media.

121 Wood, A.J., Wollenberg, B.F., and Sheblé, G.B. (2014). *Power Generation, Operation, and Control*, 3e. Wiley.

122 Seifi, H. and Sepasian, M.S. (2011). *Electric Power System Planning: Issues, Algorithms and Solutions*. Springer Science and Business Media.

2

Modeling of Energy Storage Systems for Power System Operation and Planning

2.1 Introduction

An energy storage system (ESS) can capture energy produced at one time for use at a later time. The energy can come from and release to different forms, including chemical, potential, electricity, and kinetic. In this book, we focus on the ESS that can convert electricity into other forms of energy for storage, and extract the electrical energy back when needed. To study the economic dispatch and expansion planning of a power system with the participation of ESS, the modeling of an ESS in the steady state is the fundamental work. The ESS model should accurately describe the process of energy conversion, considering the factors that limit the energy conversion comprehensively.

In present literatures, the modeling of ESSs in the steady state generally can be divided into two categories. The first category is to model for a particular type of ESS. Reference [1] studies the single cell and equivalent circuit model of a lithium ion battery. The modeling of an all-vanadium redox flow battery is studied in reference [2]. In reference [3], the research aims to model the ultracapacitor and simulation verifications are carried out. This type of modeling is to establish a detailed model based on the underlying principle of a particular form of ESS. The model can be used for improved product design, state estimation (e.g. estimating the state of charge), and optimal control of the specific ESS. The second category is to model the "energy storage system" as a whole with some simplifications, e.g. assuming the output power of the ESS is continuously adjustable, neglecting the energy conversion efficiency of the storage system and so on. In reference [4], the definition of an energy storage device in a power grid is given first. In order to assess the operational value of ESS better using different technologies and market operations, an operational value index of ESS in the electricity market is then proposed. Reference [5] proposes a general model of ideal and generic ESS with some assumptions for simplicity. These assumptions include: there are no up or down ramping limits;

Energy Storage for Power System Planning and Operation, First Edition. Zechun Hu.
© 2020 John Wiley & Sons Singapore Pte. Ltd.
Published 2020 by John Wiley & Sons Singapore Pte. Ltd.

there is no hysteresis in loading or discharging; there are no stored energy losses; and so on. The aim of the work in reference [6] is to establish a generic model of different types of ESSs. The proposed steady-state model for a power flow calculation is simple, while the proposed dynamic model is a combination of the steady state model and the power output control logics for ESS.

This chapter will first analyze the principles of a pumped hydroelectric storage system, a battery storage system, and a compressed air storage system, and then derive the generic steady-state model of ESSs. This generic steady-state model can be used in the optimal operation and planning of a power system with participation of different types of large-scale ESSs.

2.2 Pumped Hydroelectric Storage System

2.2.1 Operation of a Pumped Hydroelectric Storage System

The pumped hydroelectric storage (PHS) system makes use of the gravitational potential energy of water, which is currently the most mature energy storage technology with the largest scale. A PHS plant typically includes a higher reservoir (forebay), a lower reservoir (afterbay), a penstock, a tail race, a surge tank, and a reversible pump turbine(s), etc. The operation of PHS has two modes: pumping water to store energy and releasing water to generate energy. The schematic diagram of a PHS plant is shown in Figure 2.1. While pumping, the water is

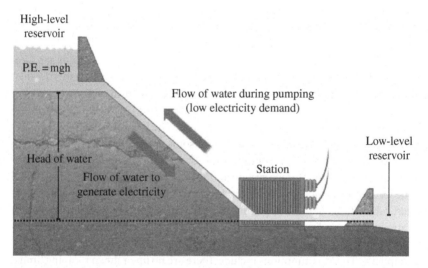

Figure 2.1 Schematic diagram of pumped hydroelectric storage system [7].

pumped through the pumper from the lower reservoir to the higher reservoir and the plant acts as an electric load and consumes energy. While releasing, the water is allowed to flow back through the turbine from the higher reservoir to the lower reservoir. When the PHS was in its early stages of development, the pumper and the turbine were separate. With technology improvements, a machine called a reversible pump turbine is now widely used, which contains both the pumper and the turbine. A reversible pump turbine can operate in two directions: it is driven by a motor to reverse as a pumper while pumping and switched into a turbine to drive the generator while releasing.

The water volumes in the higher and lower reservoirs of a PHS plant are adjusted by the reversible pump turbines. Moreover, the water volumes in the reservoirs are also influenced by the rainfall and evaporation, etc. A reservoir must have measures to deal with floods and overpumping, such as a spillway to allow the overflow water to avoid damaging the reservoir [8].

2.2.2 Steady-State Model of a Pumped Hydroelectric Storage System

The operation parameters and variables of the pumped hydroelectric storage system are summarized in Table 2.1. It can be seen that the majority of operation parameters and variables are hydraulic quantities.

According to the operational requirements of the reversible pump turbine, the pumping power p_t and active power of generator g_t in the pumped hydroelectric storage system at time t are subject to

$$u_t^{\mathrm{p}} \cdot p^{\min} \leq p_t \leq u_t^{\mathrm{p}} \cdot p^{\max} \tag{2.1}$$

$$u_t^{\mathrm{g}} \cdot g^{\min} \leq g_t \leq u_t^{\mathrm{g}} \cdot g^{\max} \tag{2.2}$$

where $p^{\min}$ and $p^{\max}$ are the lower and upper limits of the pumping power, respectively; $g^{\min}$ and $g^{\max}$ are the lower and upper limits of active power of the generator; with state control variables $u_t^{\mathrm{p}}, u_t^{\mathrm{g}} \in \{0, 1\}$.

In general, a pumped hydroelectric storage system should stop pumping while generating electricity and stop releasing water while consuming electricity by pumping. Hence, the state control variables are subject to

$$0 \leq u_t^{\mathrm{p}} + u_t^{\mathrm{g}} \leq 1 \tag{2.3}$$

Constraint (2.3) shows that the pumped hydroelectric storage system can only be pumping, generating, or in an idle state.

The ramping rate of a pumped hydroelectric storage system is high, which could reach the maximum output from starting generation within one minute. Hence, the ramping rate can be neglected in an economic dispatch with the time interval of 5 or 15 minutes.

Table 2.1 Operational parameters and variables of a pumped hydroelectric storage system.

Type	Units	Pumping mode	Releasing mode
Operation parameters	m	Lift H	Hydraulic head H
	%	Pumping efficiency η^p	Generating efficiency η^g
	MW	Upper and lower limits of pumping power p^{max}, p^{min}	Upper and lower limits of generating power g^{max}, g^{min}
	$	Start-up/shut-down costs of each pumper c^{pu}, c^{pd}	Start-up/shut-down costs of each turbine c^{tu}, c^{td}
	m^3	Inflow, including rainfall, in higher reservoir and lower reservoir^a $V_t^{u,in}$, $V_t^{d,in}$	
	m^3	Outflow, including evaporation, in higher reservoir and lower reservoir^a $V_t^{u,out}$, $V_t^{d,out}$	
	m^3	Maximum/minimum capacities of higher reservoir $V^{u,max}$, $V^{u,min}$	
	m^3	Maximum/minimum capacities of lower reservoir $V^{d,max}$, $V^{d,min}$	
Variables	m^3/s	Pumping water flow F_t^u	Pouring water flow F_t^d
	MW	Consuming power p_t	Generating power g_t
	/	Pumping state variable u_t^p	Releasing state variable u_t^g
	m^3	Volumes of water in higher and lower reservoirs V_t^u, V_t^d	

^{a} Not including the water used for pumping or generating.

To reduce the mechanical wear of a pumped hydroelectric storage system, the maximum number of state switchings should be limited:

$$\sum_{t=1}^{T} \left(\left| u_t^p - u_{t-1}^p \right| + \left| g_t^p - g_{t-1}^p \right| \right) \leq N \tag{2.4}$$

where T denotes the number of time intervals in an operational cycle (e.g. 24 hours) and N denotes the maximum allowable number of state switchings.

For a pumped hydroelectric storage plant, the residual water volumes in the higher and lower reservoirs between two successive time intervals can be calculated as follows:

$$V_t^u = V_{t-1}^u + V_t^{u,in} - V_t^{u,out} + F^u(p_{t-1}, H) \cdot \Delta t - F^d(g_{t-1}, H) \cdot \Delta t \tag{2.5}$$

$$V_t^d = V_{t-1}^d + V_t^{d,in} - V_t^{d,out} - F^u(p_{t-1}, H) \cdot \Delta t + F^d(g_{t-1}, H) \cdot \Delta t \tag{2.6}$$

where $F^u(p_t, H)$ denotes the flow rate of pumping water from the lower reservoir to the higher reservoir, which is determined by the power of the pumper and lift height, and $F^d(g_t, H)$ denotes the flow rate of releasing water from the higher reservoir to the lower reservoir, which is determined by the power of the turbine and hydraulic height. For simplicity, the functions $F^u(p_t, H)$ and $F^d(g_t, H)$ can be transformed into linear functions by piecewise linearization [9]. V_t^u and V_t^d are the water volumes in the higher and the lower reservoirs at the beginning of time interval t, respectively, $V_t^{u,in}$ and $V_t^{d,in}$ are the inflow in the higher and the lower reservoirs, respectively, while $V_t^{u,out}$ and $V_t^{d,out}$ are the outflow from the corresponding reservoirs, respectively. Equations (2.5) and (2.6) are the balance functions of water volumes in the higher and the lower reservoirs. **It should be noted that the status, p_t and g_t of each reversible turbine should be considered in all the above constraints for a PHS plant with multiple turbines.**

The inventory quantities of water are also limited by the capacity of each reservoir:

$$V^{u,min} \leq V_t^u \leq V^{u,max} \tag{2.7}$$

$$V^{d,min} \leq V_t^d \leq V^{d,max} \tag{2.8}$$

where $V^{u,max}$ and $V^{u,min}$ denote the maximum and minimum volumes of the higher reservoir, respectively, and $V^{d,max}$ and $V^{d,min}$ denote the maximum and minimum volumes of the lower reservoir, respectively.

2.3 Battery Energy Storage System

2.3.1 Operation of a Battery Energy Storage System

A battery energy storage system (BESS) consists of battery packs, power converters, power control system, and other components like protection devices, etc. Since the voltage, power, and energy of a battery cell are not enough for integrating into a power system, battery cells are generally assembled into battery packs by series–parallel connection. The conversion between electrical energy and chemical energy is achieved by the redox reaction of an anode and a cathode in the battery cells. The reaction principles of five battery systems while discharging are shown in Table 2.2. These five kinds of BESS can be used for large-scale applications. For a packaged battery cell, the anode, cathode, and electrolyte are all sealed in the cell. For a flow battery, the anode and cathode are sealed in the cell, whereas the electrolyte is stored in two separate containers. The schematic diagrams of a packaged battery cell and a flow battery are shown as Figures 2.2 and 2.3 [10].

According to the direction of energy flow between the BESS and the power network, the status of a BESS includes three modes: charging, discharging, and idle. In the idle mode, the battery storage system usually losses energy slowly due to its internal chemical reaction, which is called self-discharge. The self-discharge of a sealed battery, e.g. a lithium-ion battery, is generally caused by internal impurities

Table 2.2 Reaction principle of five battery energy storage systems while discharging.

Battery type	Cathode material	Anode material	Electrolyte	Discharging reaction
Lead–acid	Pb	PbO_2	H_2SO_4	Cathode: $Pb + SO_4^{2-} \leftarrow \rightarrow PbSO_4 + 2e^-$ Anode: $PbO_2 + 4H^+ + SO_4^{2-} + 2e^- \leftarrow \rightarrow PbSO_4 + H_2O$
Sodium–sulfur	Na	S	β-Al_2O_3	Cathode: $2Na \leftarrow \rightarrow 2Na^+ + 2e^-$ Anode: $xS + 2e^- \leftarrow \rightarrow xS^{2-}$
Lithium-ion	C	$LiCoO_2$	Organic solvent	Cathode: $Li_xC_6 \leftarrow \rightarrow 6C + xLi^+ + xe^-$ Anode: $Li_{1-x}CoO_2 + xLi^+ + xe^- \leftarrow \rightarrow LiCoO_2$
Nickel-metal hydride	Hydrogen absorbing alloy	$Ni(OH)_2$	Alkaline solution such as KOH	Cathode: $1/2H_2 + OH^- \leftarrow \rightarrow H_2O + e^-$ Anode: $NiOOH + H_2O + e^- \leftarrow \rightarrow Ni(OH)_2 + OH^-$
Vanadium redox	$V^{2+} \leftarrow \rightarrow V^{3+}$	$V^{5+} \leftarrow \rightarrow V^{4+}$	Dilute H_2SO_4	Cathode: $V^{2+} \leftarrow \rightarrow V^{3+} + e^-$ Anode: $V^{5+} + 2H^+ + e^- \leftarrow \rightarrow V^{4+} + H_2O$

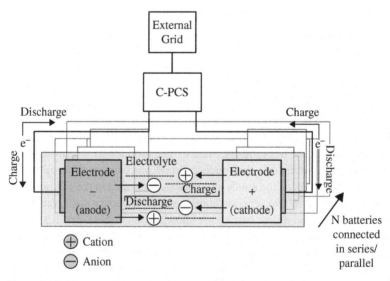

Figure 2.2 Schematic diagram of a packaged battery cell [10].

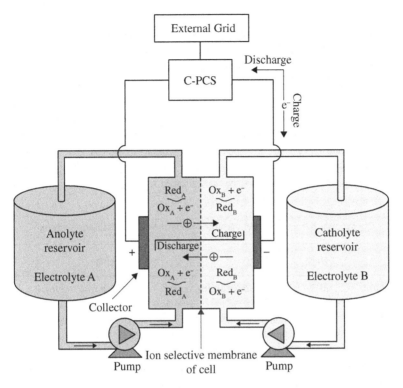

Figure 2.3 Schematic diagram of a flow battery system [10].

of the battery, self-fluxing of an electrode, accumulation of an electrolyte or grease on the battery cover and a short-circuit within the battery. The self-discharge of a flow battery happens when electrolyte flows through the battery, generally caused by the concentration reduction of effective reactants resulting from ion penetration and chemical reaction when electrolytes flow through the amberplex [11]. The self-discharge of a battery is measured by the self-discharge rate, which is defined as the ratio of energy loss to the battery capacity per month at a particular temperature, generally calibrated at 20 °C [12]. The self-discharge rate of a battery is related to the battery type and materials. The typical self-discharge rates of five BESSs are listed in Table 2.3 [10, 12].

2.3.2 Steady-State Model of a Battery Energy Storage System

For a BESS, its battery packs connect with the power grid through converters. The power exchange between battery packs and the power system can be tuned by controlling the converters. Hence, the operational parameters and variables are

Table 2.3 Typical self-discharge rate of five battery energy storage systems.

Battery type	Self-discharge rate (%)
Lead–acid	5
Sodium–sulfur	<1
Lithium-ion	5~10
Nickel-metal hydride	>30
All-vanadium redox	<1

Table 2.4 Operational parameters and variables of battery energy storage system.

Type	Units	Charging	Discharging
Operational parameters	MW	Minimum charging power $p^{c,\min}$	Minimum discharging power $p^{d,\min}$
	MW	Maximum discharging power $p^{c,\max}$	Maximum discharging power $p^{d,\max}$
	%	Charging efficiency η^c	Discharging efficiency η^d
	MWh	Maximum energy capacity $E^{\max}$	
	MWh	Minimum residue energy capacity $E^{\min}$	
	%	Self-discharging rate ξ	
Variables	MW	Charging power p_t^c	Discharging power p_t^d
	/	Binary variable of charging state u_t^c	Binary variable of discharging state u_t^d
	MWh	Stored energy E_t	

mainly electromagnetic quantities, as shown in Table 2.4. These parameters and variables can be directly used in the mathematical formulations for power system optimal planning and dispatch.

Limited by both battery and converter capacities, the charging power p_t^c and discharging power p_t^d of a BESS are subject to

$$u_t^c \cdot p^{c,\min} \le p_t^c \le u_t^c \cdot p^{c,\max} \tag{2.9}$$

$$u_t^d \cdot p^{d,\min} \le p_t^d \le u_t^d \cdot p^{d,\max} \tag{2.10}$$

where the binary variables of the BESS state are defined as

$$u_t^c = \begin{cases} 0, \text{Not charging} \\ 1, \text{Charging} \end{cases}, \quad u_t^d = \begin{cases} 0, \text{Not discharging} \\ 1, \text{Discharging} \end{cases}.$$

Similar to a PHS plant, u_t^c and u_t^d are also subject to

$$0 \leq u_t^c + u_t^d \leq 1 \tag{2.11}$$

Constraint (2.11) shows that a BESS cannot charge and discharge at the same time.

The stored energy of a battery storage system in two successive time intervals should be balanced:

$$E_t = E_{t-1} \cdot (1 - \xi) + p_{t-1}^c \cdot \Delta t \cdot \eta^c - p_{t-1}^d \cdot \Delta t / \eta^d \tag{2.12}$$

where η^c and η^d denote charging and discharging efficiencies of the BESS, respectively. The charging efficiency is related to the battery technology, battery materials, and environmental temperature, whereas the discharging efficiency is related to the discharge power, environmental temperature. and internal resistance; ξ denotes the self-discharge rate of a battery storage system, which is related to internal characteristics of the battery cells.

The stored energy of a BESS should be within the minimum and maximum energy levels:

$$E^{\min} \leq E_t \leq E^{\max} \tag{2.13}$$

where $E^{\min}$ and $E^{\max}$ denote the minimum and the maximum allowable energy capacities of BESS, respectively.

2.4 Compressed Air Energy Storage System

2.4.1 Operation of a Compressed Air Energy Storage System

The compressed air energy storage (CAES) is one of the few energy storage technologies that is suitable for long duration (up to tens of hours) and utility scale (up to hundreds to thousands of megawatts) applications [8]. A CAES consists of a compressor, combustor and heat recuperator, gas turbine, air storage reservoir, motor/generator, and so on. A CAES can be used for storing energy or generating electricity when necessary. When storing energy, electricity from the power system is used to run the motor and drive the compressor to inject air into the air storage reservoir. When generating electricity, the high-pressure air is released from the air storage reservoir and is usually combusted with the fuel in a combustor, thus releasing energy to drive the gas turbine to run the generator [13], shown in Figure 2.4. Depending on the reservoir type for air storage, a CAES can operate in a number of ways [8].

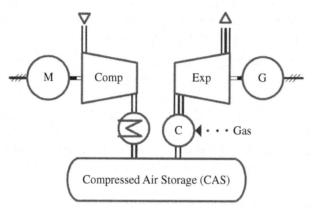

Figure 2.4 Schematic diagram of a diabatic compressed air storage system with a heat source.

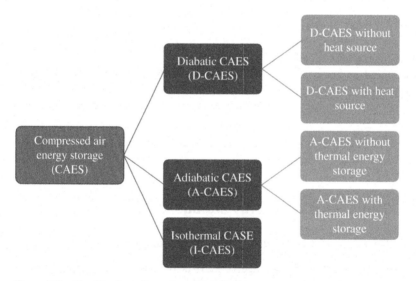

Figure 2.5 Classification of a compressed air storage system.

According to the adiabatic methods and heat sources, the CAES can be divided into different types, as shown in Figure 2.5. The main criterion for categorization is the method describing how heat is handled during compression and before expansion of the air [13].

1) Diabatic CAES (D-CAES) without a heat source. While generating power, the compressed high-pressure air drives the gas turbine directly. Although it has a simple structure and zero pollution or emission, this kind of CAES is generally

suitable for the low-power applications due to its low energy density and low efficiency.

2) D-CAES with a heat source. This type of CAES combusts fuel (usually gas) to provide heat when generating electricity. This will result in fuel consumption and emission of greenhouse gases and pollution.

3) Adiabatic CAES (A-CAES) without thermal energy storage (TES). For D-CAES, the heat resulting from air compression and air cooling is wasted. The A-CAES without TES stores the hot air inside a combined thermal energy and compressed air storage volume. However, the temperature can reach up to 277°C when adiabatically compressing ambient air just to a moderate pressure of 10 bar. Thus, A-CAES without TES cannot be expected for a commercial application in the near future due to a considerable material requirement.

4) A-CAES with TES. For A-CAES with TES, the heat energy is extracted and stored in separate TES devices before the compressed air enters the air storage reservoir (see Figure 2.6). When generating power, the compressed air and the stored heat energy are recombined and expanded through the turbine. The TES can significantly improve the overall efficiency of the CAES.

5) Isothermal CAES (I-CAES). An I-CAES tries to keep the temperature constant during the air compression and expansion processes, and does not use fuel during the expansion cycle. All the I-CAES concepts known so far are based on piston machinery.

The ramp rate of a compressed air storage system is high, e.g. the ramp rate of the McIntosh compressed air storage plant is about 18 MW min^{-1} [8]. To initiate compression operation, the gas turbine typically brings the machinery train to a certain speed. The compressors keep operating after synchronization, when the gas

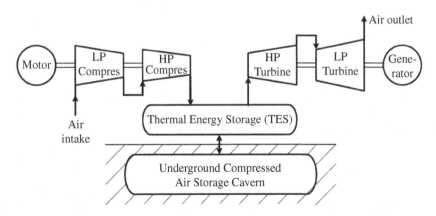

Figure 2.6 Function diagram of an A-CAES with thermal energy storage [14].

turbine is tripped and turned off. In other words, the gas turbine is used to initiate both compression and generation. Hence, it would take some time for CAES to switch from one operating mode to another, during which the compressed air storage system is neither generating power nor compressing air. The switching time of a Huntorf compressed air storage plant, in Germany, is 20 minutes. There is a risk of air leakage from the air storage reservoir of CAES, but the leakage rate should be kept below a very low level.

2.4.2 Steady-State Model of a Compressed Air Energy Storage System

According to the basic principles of CAES, the operational parameters and variables of a general CAES are listed in Table 2.5. The majority of the parameters and variables are thermodynamic quantities.

Table 2.5 Operational parameters and variables of CAES.

Type	Units	Charging	Discharging
Operational parameters	%	Charging efficiency η^p	Discharging efficiency η^g
	\$	Start-up/shut-down costs of compressor c^{pu}, c^{pd}	Start-up/shut-down costs of gas turbine c^{tu}, c^{td}
	MW	Maximum/Minimum compressing power p^{max}, p^{min}	Maximum/Minimum generating power g^{max}, g^{min}
	MW min^{-1}	/	Maximum ramp up/down rate R^u, R^d
	min	Minimum operating time while charging T^p	Minimum operating time while generating T^g
	m^3	Volume of air storage reservoir V	
	min	Required time for state switchings T^s	
	Pa	Maximum/Minimum air pressure of air storage reservoir κ^{max}, κ^{min}	
Variables	m^{3}h^{-1}	/	Consumption of oil/gas V^c
	MW	Charging power p_t	Generating power g_t
	/	Binary variable of charging state u_t^p	Binary variable of discharging state u_t^g
	/	Binary variables of start-up/shut-down state for charging ς_t^{pu}, ς_t^{pd}	Binary variables of start-up/shut-down state for discharging ς_t^{gu}, ς_t^{gd}
	Pa	Air pressure of air storage reservoir κ_t	

For a CAES, the motor power while charging p_t, the generating power while discharging g_t, and the state variables u_t^p, u_t^g should satisfy constraints (2.1) to (2.4). In the air storage reservoir, energy is stored as potential energy or, more precisely, exergy of compressed air. According to the analysis in references [14] to [16], the power of the compressor when charging can be calculated by

$$p_t = \frac{1}{\eta^c} \dot{m}_t^c c^p \theta^{\text{cin}} \left(\beta^{n^c - \frac{1}{n^c}} - 1 \right) \tag{2.14}$$

where $\dot{m}_t^c$ is the mass flow of air (kg s^{-1}) when compressing, c^p is the specific heat capacity at constant pressure (J kg·K^{-1}), θ^{cin} stands for the air temperature at the inlet of the compressor, β is the compression ratio, which is decided by the air pressure after and before compression, and n^c is the polytropic exponent that can be obtained from the polytropic efficiency of the compressor [15, 16].

In a similar way, the generating power of the turbine when discharging can be calculated by

$$g_t = \eta^g \dot{m}_t^d c^p \theta^{\text{din}} \left(1 - \beta^{n^d - \frac{1}{n^d}} \right) \tag{2.15}$$

where $\dot{m}_t^d$ is the mass flow of air (kg s^{-1}) when expanding, θ^{din} stands for the air temperature at the inlet of the turbine, β is the compression ratio of the turbine, and n^d is the polytropic exponent of the turbine.

For simplicity, only one stage of compression and expansion for the CAES is discussed here and the temperature variation within the air storage reservoir (cylinder or cavern) is neglected. We consider that the volume of the air storage reservoir is constant. Then the pressure change of the stored air can be represented as

$$\kappa_t = \kappa_{t-1} + \frac{R^a \theta^r}{V} \left(\dot{m}_t^c - \dot{m}_t^d \right) \Delta t \tag{2.16}$$

where R^a is the gas constant for dry air (J kg·K^{-1}), θ^r and V are the air temperature (K) and the volume (m^3) of the air storage reservoir, respectively, and Δt is the duration of each time interval. This equation represents the balance of pressure within the air storage reservoir. The pressure of the air storage reservoir should be subject to

$$\kappa^{\min} \le \kappa_t \le \kappa^{\max} \tag{2.17}$$

The generating power of a CAES is subject to the limitations of the ramp rate:

$$g_t - g_{t-1} \le u_{t-1}^g \cdot R^u \tag{2.18}$$

$$g_t - g_{t-1} \ge -u_t^g \cdot R^d \tag{2.19}$$

To avoid frequent state switchings, the compressed air storage system is subject to the minimum state holding time:

$$T^{\mathrm{p}} \cdot \varsigma_t^{\mathrm{pu}} \leq \sum_{i=t}^{t+T^c-1} u_i^{\mathrm{p}} \leq T^{\mathrm{p}} \tag{2.20}$$

$$T^{\mathrm{g}} \cdot \varsigma_t^{\mathrm{gu}} \leq \sum_{i=t}^{t+T^{\mathrm{g}}-1} u_i^{\mathrm{g}} \leq T^{\mathrm{g}} \tag{2.21}$$

$$T^{\mathrm{s}} \cdot \left(\left(\varsigma_t^{\mathrm{pd}} \geq 1 \right) \mid \left(\varsigma_t^{\mathrm{gd}} \geq 1 \right) \right) \leq \sum_{i=t}^{t+T^{\mathrm{s}}-1} (u_i^{\mathrm{p}} \leq 0)(u_i^{\mathrm{g}} \leq 0) \leq T^{\mathrm{s}} \tag{2.22}$$

Constraints (2.20) and (2.21) limit the minimum duration time that a CAES should stay in a charging or discharging state. Constraint (2.22) represents the case where the CAES should be neither generating power nor compressing air for a fixed time T^s during state switchings.

The D-CAES with a heat source needs to consume fuel to provide heat while generating power. If the fuel is natural gas, for example, the consumption rate ($\mathrm{m}^3\ \mathrm{h}^{-1}$) of the gas in the combustor can be calculated as follows:

$$V_t^c = b \cdot \beta \cdot \frac{g_t \cdot LHV}{HV_g} \tag{2.23}$$

where *LHV* is the heat rate of natural gas (the fuel energy consumed for per unit of effective power), HV_g is the lower calorific value of natural gas, and *b* is a coefficient related to the type of CAES:

$$b = \begin{cases} 0, & \text{nonadiabatic CAES without a heat source} \\ & \text{or adiabatic CAES with heat storage} \\ 1, & \text{nonadiabatic CAES which combusts fuel to provide heat} \end{cases} \tag{2.24}$$

where $\beta(\beta < 1)$ represents the fuel saved coefficient by an exhaust-heat recovery device, valued according to the specific situation.

2.5 Simplified Steady-State Model of a Generic Energy Storage System

From the steady-state models discussed above for the three different ESSs, it can be seen that the model varies with the type of storage system, due to the different

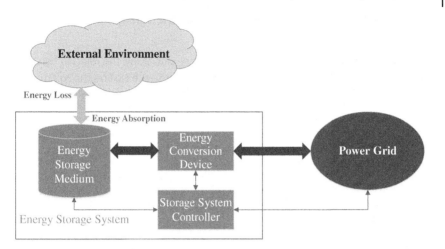

Figure 2.7 Schematic diagram of the characteristics of the storage system.

energy storage principles. However, some common characteristics of the storage systems and their models can be summarized as follows (shown as Figure 2.7):

1) They mainly consist of an energy storage device, energy conversion device, and controller.
2) The maximum stored energy in an ESS is limited by its storage device and the level of stored energy is controllable.
3) The energy storage device may exchange energy (or energy storage media) with the environment, acquire energy from the environment (e.g. a PHS plant can acquire energy from rainfall) and lose energy to the environment (e.g. a PHS plant can lose energy because of evaporation).
4) The energy conversion device can bidirectionally exchange energy with a power system and absorb electrical energy from the power system while charging and inject electrical energy to power system while discharging.
5) The exchanged power is usually controllable, which mainly depends on the parameters of energy storage device and energy conversion device.
6) The energy conversion device may get external energy during energy conversion, e.g. the D-CAES need to combust fuel for power generation.
7) The processes of charging and discharging are not 100% efficient; that is to say, there are energy losses.

Based on the above characteristics, we can transform the operational parameters and constraints of different ESSs into electrical quantities. Then the specific energy conversion process can be neglected and the simplified steady-state model can be derived for a generic storage system.

2.5.1 Transformation of a Pumped Hydroelectric Storage System Model

According to the principles of the PHS system, we can derive the relation between the mechanical quantities and electromagnetic quantities based on the equation of gravitational potential energy:

$$p_t \cdot \eta^p = \rho \cdot \gamma \cdot F^u(p_t, H) \cdot H \tag{2.25}$$

$$g_t = \rho \cdot \gamma \cdot F^d(g_k, H) \cdot H \cdot \eta^g \tag{2.26}$$

$$E_t = \rho \times \gamma \times H \times V_t^u \tag{2.27}$$

where ρ is the inventory water density, valued at $1.0 \times 10^3 \text{kg m}^{-3}$ for fresh water and $1.025 \times 10^3 \text{kg m}^{-3}$ for seawater and γ is the gravitational acceleration and its valued is commonly set as 9.8 m s^{-2}.

The constraints (2.5) to (2.8) can be changed into the following forms using (2.25) to (2.27):

$$E_t = E_{t-1} + \Delta E_{t-1}^{ext} + p_{t-1} \cdot \Delta t \cdot \eta^c - g_{t-1} \cdot \Delta t / \eta^d \tag{2.28}$$

$$E^{\min} \leq E_t \leq E^{\max} \tag{2.29}$$

where

$$\Delta E_{t-1}^{ext} = \rho \gamma H \left(V_t^{u,in} - V_t^{u,out} - V_t^{u,pro} \right) \tag{2.30}$$

$$E^{\max} = \rho \gamma H V_t^{u,\max} \tag{2.31}$$

$$E^{\min} = \rho \gamma H V_t^{u,\min} \tag{2.32}$$

where ΔE_t^{ext} is the total energy exchange of the PHS system with the environment at time t, $\Delta E_t^{ext} > 0$ indicates that the PHS gets energy from the environment, and $\Delta E_t^{ext} < 0$ indicates that the PHS loses energy to the environment. By the above transformation, the constraints of the model of a pumped hydroelectric storage system can be formulated by the electromagnetic quantities $p_t, g_t, \Delta E_t^{ext}$, and E_t.

2.5.2 Transformation of a Compressed Air Energy Storage System Model

According to the theory of pneumatics, the available energy of compressed air is determined by the pressure, volume, and temperature of air, which can be mathematically formulated as follows [17]:

$$E = \kappa V \left[\ln \frac{\kappa}{\kappa^a} + \frac{\alpha}{\alpha - 1} \left(\frac{\theta^r - \theta^a}{\theta^a} - \ln \frac{\theta^r}{\theta^a} \right) \right] \tag{2.33}$$

where κ^a is the absolute pressure of the atmosphere, θ is the absolute temperature within the air reservoir, θ^a is the absolute temperature of the atmosphere, and α is the specific heat ratio of the air. Considering the minimum and maximum air pressure requirements of the air storage reservoir, the energy limits of a CAES can be obtained by the following equations:

$$E^{\min} = \kappa^{\min} V \left[\ln \frac{\kappa^{\min}}{\kappa^a} + \frac{\alpha}{\alpha - 1} \left(\frac{\theta^r - \theta^a}{\theta^a} - \ln \frac{\theta^r}{\theta^a} \right) \right] \tag{2.34}$$

$$E^{\max} = \kappa^{\max} V \left[\ln \frac{\kappa^{\max}}{\kappa^a} + \frac{\alpha}{\alpha - 1} \left(\frac{\theta^r - \theta^a}{\theta^a} - \ln \frac{\theta^r}{\theta^a} \right) \right] \tag{2.35}$$

The constraints contain thermodynamic quantities, which can all be changed into the electromagnetic quantities.

2.5.3 Steady-State Model of a Generic Energy Storage System

For a battery storage system with a self-discharging rate ξ, if we define

$$\Delta E_t^{ls} = \xi E_t \tag{2.36}$$

then Eq. (2.12) can be rewritten as

$$E_t = E_{t-1} - \Delta E_{t-1}^{ls} + p_{t-1}^c \cdot \Delta t \cdot \eta^c - p_{t-1}^d \cdot \Delta t / \eta^d \tag{2.37}$$

where ΔE_t^{ls} denotes the energy losses within the time interval Δt.

By the above transformations, the generic steady-state model for the PHS, CAES, and BESS can be formulated by the parameters and variables listed in Table 2.6. The operational parameters include the system capacity range, the charging/discharging power range, the maximum number of state switchings, energy exchange with the external environment, energy loss, maximum ramp up/down rate, the minimum state holding time while storing and releasing, the fixed time for state switchings, and operation efficiency. The output variables include power, state, stored energy, the number of state switchings, and the power absorbed from the environment at each interval.

In the generic steady-state model of ESS, the constraints include power limit constraints, the energy balance between adjacent time intervals, the mutual exclusion of state variables, constraints for the number of state switchings, the ramp rate and stored energy constraints, etc.

Charging power limits:

$$u_t^c \cdot p^{c,\min} \leq p_t^c \leq u_t^c \cdot p^{c,\max} \tag{2.38}$$

Table 2.6 Parameters of simplified steady-state model of a generic energy storage system.

	Description	Symbol/expression
Operational parameter	Minimum residue capacity and maximum capacity	$E^{\min}, E^{\max}$
	Minimum and maximum storing power	$p^{c,\min}, p^{c,\max}$
	Minimum and maximum releasing power	$p^{d,\min}, p^{d,\max}$
	Maximum number of state switchings during a specified operational period (e.g. one day)	N
	Exchanged energy with external environment in a time interval	ΔE_t^{ex}
	Energy loss in a time interval	ΔE_t^{ls}
	Maximum ramp up/down rate	R^u, R^d
	Minimum holding time while charging/discharging	τ^c, τ^d
	Fixed time for state switchings	τ^s
	Efficiency of charging/discharging	η^c, η^d
	Binary variables of start-up/shut-down state for charging	$\varsigma_t^{cu}, \varsigma_t^{cd}$
	Binary variables of start-up/shut-down state for discharging	$\varsigma_t^{du}, \varsigma_t^{dd}$
Output variables	Power of charging/discharging	p_t^c, p_t^d
	Binary state variable of charging/discharging	$u_t^c, u_t^d \in \{0,1\}$
	Stored energy	E_t

Discharging power limits:

$$u_t^d \cdot p^{d,\min} \le p_t^d \le u_t^d \cdot p^{d,\max} \tag{2.39}$$

Constraint of energy balance between adjacent time intervals:

$$E_t = E_{t-1} - \Delta E_{t-1}^{ls} + \Delta E_{t-1}^{ab} + p_{t-1}^c \cdot \Delta t \cdot \eta^c - p_{t-1}^d \cdot \Delta t / \eta^d \tag{2.40}$$

Mutual exclusion of state variables:

$$0 \le u_t^c + u_t^d \le 1 \tag{2.41}$$

Constraint for the number of state switchings:

$$\sum_{t=1}^{T} \left(\left| u_t^c - u_{t-1}^c \right| + \left| u_t^d - u_{t-1}^d \right| \right) \le N \tag{2.42}$$

Ramp rate constraints:

$$p_t^d - p_{t-1}^d \leq u_{t-1}^d \cdot R^u \tag{2.43}$$

$$p_t^d - p_{t-1}^d \geq -u_t^d \cdot R^d \tag{2.44}$$

Limits of residue energy:

$$E^{\min} \leq E_t \leq E^{\max} \tag{2.45}$$

Constraint for the minimum state holding time if $\tau^c > \Delta t$ or $\tau^d > \Delta t$:

$$\tau^c \cdot \varsigma_t^{cu} \leq \sum_{i=t}^{t+\tau^c-1} u_i^c \leq \tau^c \tag{2.46}$$

$$\tau^d \cdot \varsigma_t^{du} \leq \sum_{i=t}^{t+\tau^d-1} u_i^d \leq \tau^d \tag{2.47}$$

Constraint for the minimum shutting down time if $\tau^s > \Delta t$:

$$\tau^s \cdot \left((\varsigma_t^{cd} \geq 1) \mid (\varsigma_t^{dd} \geq 1) \right) \leq \sum_{i=t}^{t+\tau^s-1} \left(u_i^c \leq 0 \right) \left(u_i^d \leq 0 \right) \leq \tau^s \tag{2.48}$$

The simplified steady-state model of a generic ESS considers the key characteristics of different ESSs. Using this steady-state model, the mathematical formulations for different power system planning and dispatch problems can be easily built to take different ESSs into account. The parameters of the simplified steady-state model should be derived from the primitive parameters of the corresponding ESS.

2.6 Conclusion

This chapter first analyses the operational mechanisms of PHS, BESS, and CAES. A simplified steady-state model for generic ESSs is then established by transforming the non-electrical parameters and variables of different types of ESSs into electrical ones. This simplified steady-state model can be used for different power system planning and operation problems considering multi-types of ESSs, which will be introduced in the following chapters.

References

1 Hussein, A.A. and Batarseh, I. (2011). An overview of generic battery models. *Power and Energy Society General Meeting*. IEEE, pp. 1–6.

2 Binyu, X., Jiyun, Z., and Jinbin, L. (2013). Modeling of an all-vanadium redox flow battery and optimization of flow rates. *Power and Energy Society General Meeting (PES)*. IEEE, pp. 1–5.

3 Grbović, P.J., Delarue, P., and Le Moigne, P. (2014). Modeling and control of ultra-capacitor based energy storage and power conversion system. *Proceedings of the IEEE 15th Workshop on Control and Modeling for Power Electronics*, Santander, Spain. IEEE.

4 Thatte, A.A. and Xie, L. (September 2012). Towards a unified operational value index of energy storage in smart grid environment. *IEEE Transactions on Smart Grid* 3 (3): 1418–1426.

5 Pozo, D., Contreras, J., and Sauma, E.E. (2014). Unit commitment with ideal and generic energy storage units. *IEEE Transactions on Power Systems* 29 (6): 2974–2984.

6 Hartmann, B. and Liptak, S. (2015). Development of steady state and dynamic energy storage models for DIgSILENT PowerFactory. *PowerTech*, Eindhoven. IEEE, pp. 1–6.

7 Pumped storage hydroelectric power station. www.bbc.co.uk/bitesize/standard/ physics/energy_matters/generation_of_electricity/revision/3.

8 Barnes, F.S. and Levine, J.G. (2011). *Large Energy Storage Systems Handbook*. Boca Raton, FL, USA: CRC Press.

9 Ding, H., Hu, Z., and Song, Y. (2012). Stochastic optimization of the daily operation of wind farm and pumped-hydro-storage plant. *Renewable Energy* 48: 571–578.

10 Díaz-González, F., Sumper, A., Gomis-Bellmunt, O. et al. (2012). A review of energy storage technologies for wind power applications. *Renewable and Sustainable Energy Reviews* 16 (4): 2154–2171.

11 Bei, L., Jian-bo, G., Dong, H., and Ji-zhong, C. (2009). Efficiency analysis of redox flow battery applied in power system. *Proceedings of the Chinese Society for Electrical Engineering* 29 (35): 1–6.

12 Brunet, Y. (2011). *Energy Storage*. Hoboken, NJ, USA: Wiley.

13 Budt, M., Wolf, D., Span, R., and Yan, J. (2016). A review on compressed air energy storage: basic principles, past milestones and recent developments. *Applied Energy* 170: 250–268.

14 Steta, F. de S. (October 2010). *Modeling of an Advanced Adiabatic Compressed Air Energy Storage (AA-CAES) Unit and an Optimal Model-Based Operation Strategy for Its Integration into Power Markets*. Swiss Federal Institute of Technology (ETH), Zurich.

15 Hartmann, N., Vöhringer, O., Kruck, C. et al. (2012). Simulation and analysis of different adiabatic compressed air energy storage plant configurations. *Applied Energy* 93: 541–548.

16 Yang, Z., Wang, Z., Ran, P. et al. (2014). Thermodynamic analysis of a hybrid thermal-compressed air energy storage system for the integration of wind power. *Applied Thermal Engineering* 66 (1): 519–527.

17 Cai, M., Kawashima, K., and Kagawa, T. (2006). Power assessment of flowing compressed air. *Journal of Fluids Engineering* 128 (2): 402–405.

3

Day-Ahead Schedule and Bid for a Renewable Energy Generation and Energy Storage System Union

3.1 Introduction

In recent years, much effort has been made around the world to develop renewable energy such as wind and solar energy in response to fossil fuel depletion and environmental problems. However, the uncertainty and fluctuation of wind and solar power have negative effects on secure and stable operation of power systems, which restrains the increase of wind and solar power penetration. Furthermore, the relatively large forecast error has a negative impact on the revenue of renewable generation in electricity markets.

In order to better accommodate wind and solar power, studies have focused on the coordination of energy storage systems (ESS) with renewable energy generation (REG). REG and ESS can form a union, which means they can offer in the day-ahead stage and get paid as a virtual unit or plant. In this way, their joint power output is more controllable, and it will offset the negative influence caused by forecast errors of REG on power systems. In reference [1], the benefits of the coordination of thermal power plants, wind farms, and an ESS on the system are studied under different seasons, with various power demands and wind power penetrations. The authors of reference [2] formulate a two-stage stochastic optimization model to maximize the profit of a wind-storage union under the Spanish power market environment, and the case studies demonstrate that the profit of coordinated operation is much higher than that of uncoordinated operation. Optimal sizing of an ESS has been studied in reference [3] to coordinate with an existing wind farm. References [4] and [5] studied the improvement that the ESS can bring to REG, power transfer capability, and power system stability.

The limited accuracy of renewable generation forecasts results in revenue loss for renewable power producers because they face risks in real-time or balancing markets to settle their deviations between day-ahead offers and actual power outputs. ESS can coordinate with REG in two ways. In one way, because ESS can be used to make arbitrage with variable electricity prices, the overall profit of WF and

Energy Storage for Power System Planning and Operation, First Edition. Zechun Hu.
© 2020 John Wiley & Sons Singapore Pte. Ltd.
Published 2020 by John Wiley & Sons Singapore Pte. Ltd.

ESS will increase in day-ahead markets compared with the situation where they are scheduled separately. In the other way, ESS can be used to flatten the fluctuations of REG output in real-time operation. A *filter operation strategy* is stated in reference [6], which utilizes ESS to compensate for the imbalance from day-ahead offering [7]. Similarly, the so-called *reserve operation strategy* is proposed in reference [8], which uses the *expected utility maximization* strategy at the offering stage but then sets contracts for ESS to cover the shortfalls of a wind farm. The work in reference [9] shows that the wind farm and ESS can both be better off through contracts for compensation of wind power deviations by ESS. The concavity of objective functions based on the greedy control strategy of ESS is studied. Although the potential arbitrage strategy of ESS is overlooked, it certainly influences day-ahead offerings and resulting market revenues.

In this chapter, we will discuss the optimization formulations that the REG–ESS union can use for making its day-ahead optimal dispatch or bidding decisions.

3.2 Basic Model for Day-Ahead Schedule of a REG–ESS Union

The objective of day-ahead schedule for the REG–ESS union is to maximize its total revenue. Because of the energy balance constraint of the ESS, it is necessary to make its charging/discharging plan for coordination with the REG and price arbitrage. It is assumed in this chapter that the REG–ESS union is a price-taker participating in the power markets. Therefore, the basic model for a day-ahead schedule of the REG–ESS union is to dispatch the ESS properly according to the forecasted REG power output and electricity prices. Taking the startup and shutdown costs of the ESS and the penalty on power deviation into account, the objective function can be formulated as

$$\max \sum_{t \in T} \left\{ \pi_t \left(p_t^{sd} + p_t^{rg} - p_t^{sc} \right) \Delta t - \omega \pi_t \left| p_t^{sd} + p_t^{rg} - p_t^{sc} - \hat{p}_t \right| \Delta t - \left(C^{su} n_t^{su} + C^{sd} n_t^{sd} \right) \right\},$$

$$t \in T \tag{3.1}$$

where π_t is the forecasted electricity price at time interval t, Δt is the duration of each time interval, and p_t^{rg} is the power output of REG. With regard to the parameters of ESS, p_t^{sd} and p_t^{sc} are the discharging and charging powers at interval t, respectively. The number of state transitions of ESS are counted to calculate the related costs, n_t^{su}, n_t^{sd} stand for the startup and shutdown times, respectively, C^{su} and C^{sd} are the startup and shutdown costs of ESS, respectively, and $\hat{p}_t$ is the scheduled power output of the REG–ESS union at interval t, which should be determined by the optimization model.

The first part in the objective function is the revenue from selling electricity, the second part is the penalty to the power deviation, and the third part is the startup and shutdown costs of the ESS. The startup and shutdown costs vary with different energy storage technologies. Generally, the startup and shutdown costs are non-negligible for storage systems such as a PHS plant and a CAES system.

The constraints that should be considered are mainly for the ESS. The constraints that are derived for a generic ESS in Chapter 2 can be used here directly.

In order to limit the deviation of the net power output of the REG–ESS union, the following constraint can be added:

$$(1 - \delta)\hat{p}_t \le p_t^{rg} + p_t^{sd} - p_t^{sc} \le (1 + \delta)\hat{p}_t, \qquad t \in T \tag{3.2}$$

where δ is the maximum allowable deviation factor. Deviation constraints and a penalty function are combined to seek the solution that can take the system operation requirement and the union's profit into consideration simultaneously.

If we further consider that the output of REG can be reduced or curtailed, then the following constraint can be added:

$$0 \le p_t^{rg} \le p_t^{rf}, \qquad t \in T \tag{3.3}$$

where p_t^{rf} means the forecast power output capability of REG.

3.3 Stochastic Optimization for Day-Ahead Coordination

Although the precisions of wind and solar power forecast have been improved significantly in recent years, the day-ahead wind or solar power forecast errors are still as large as 10–20% at present [10, 11]. To consider the influence that uncertainties and forecast errors of renewable power generation have on the day-ahead decision, scenario-based and chance-constrained stochastic optimization models are given as follows.

3.3.1 Scenario-Based Optimization Model

The uncertain REG output can be depicted by scenarios and each scenario represents a forecast of a generation power profile within the concerned time horizon. For the day-ahead dispatch of the REG–ESS union, the objective function of the scenario-based optimization model can be written as:

$$\max \sum_{s=1}^{S} \rho_s \sum_{t \in T} \left\{ \pi_t \left(p_{t,s}^{rg} + p_t^{sd} - p_t^{sc} \right) - \omega \pi_t \mid p_{t,s}^{rg} + p_t^{sd} - p_t^{sc} - \hat{p}_t \mid - \left(C^{su} n_t^{su} + C^{sd} n_t^{sd} \right) \right\}$$

$$\tag{3.4}$$

where S is the number of scenarios, ρ_s is the weight or probability of scenario s, and $p_{t,s}^{rg}$ stands for the power output of REG under scenario s. It should be noted that the charging and discharging power of ESS under all the scenarios are the same. We will consider the dynamic change of the operating state of ESS later.

Different distributions have been proposed to model the forecast error of renewable power [12]. If the forecast error of wind power is assumed to obey a normal distribution $N(\mu, \sigma^2)$, the weight of each scenario ρ_i can be determined by the following set of equations when three scenarios are considered [13]:

$$\begin{cases} \rho_1 + \rho_2 + \rho_3 = 1 \\ \rho_1 x_1 + \rho_2 x_2 + \rho_3 x_3 = \int_{-\infty}^{+\infty} f(x)x\,dx = \mu \\ \rho_1 x_1^2 + \rho_2 x_2^2 + \rho_3 x_3^2 = \int_{-\infty}^{+\infty} f(x)x^2\,dx = \mu^2 + \sigma^2 \end{cases} \tag{3.5}$$

where x_1, x_2, and x_3 represent the scenarios of -3σ, 0, $+3\sigma$, and ρ_1, ρ_2, ρ_3 are the weights of scenarios, respectively. Assume $\mu = 0$ and $\sigma = 0.1p_t^{rf}$; solving the above equations obtain $\rho_1 = \rho_3 = 0.056$, $\rho_2 = 0.888$.

Correlations between the successive time intervals are not considered in the above method. Furthermore, when more than 10 scenarios are taken into account, analytic methods to obtain the weights of all scenarios may encounter a numerical stability problem. Considering these two important factors, simulation-based methods are usually employed by sampling based on the forecast error distribution and autocorrelation of REG output [14]. Then scenario reduction or clustering methods can be used to get an appropriate number of scenarios and the corresponding weight of each scenario.

The output deviation constraint should be taken into account for each scenario. For example, if three scenarios are considered with weighting factors obtained by Eq. (3.5), the following constraints can be set:

$$(1 - \delta)\hat{p}_t \leq p_{t,s}^{rg} + p_t^{sd} - p_t^{sc} \leq (1 + \delta)\hat{p}_t, \quad s = 2, t \in T \tag{3.6}$$

$$p_{t,s}^{rg} + \overline{p}^{sd} \geq \hat{p}_t, \quad s = 1, 3, t \in T \tag{3.7}$$

Constraint (3.6) sets a deviation constraint on the scenario with the expected wind power. Constraint (3.7) requires the REG–ESS union to have adequate reserves to meet the power output deviation under the other two extreme scenarios. In Eq. (3.7), $\overline{p}^{sd}$ is the maximum discharging power of the ESS. All the other constraints of the basic dispatch model should also be applied to this model.

3.3.2 Chance-Constrained Optimization Model

Chance-constrained programming is another type of stochastic optimization method that can deal with the random distributions of uncertain parameters [15]. For many optimization problems, some constraints cannot always be satisfied or the costs are very high in order to guarantee the bindings of all constraints under all possible values of the random parameters. For chance-constrained programming, the probabilities of meeting all or part of the inequality constraints are relaxed to preset levels lower than 1.0.

The objective function of a chance-constrained optimization model can be set the same as that of the deterministic one. The main difference is that Eq. (3.2) for the deterministic model is replaced by the following chance constraints with different confidence levels to limit the deviation of the actual power output:

$$Pr\left\{\widetilde{p}_t^{rg} + p_t^{sd} - p_t^{sc} \leq (1 + \xi)\hat{p}_t\right\} \geq \alpha, \qquad t \in T \tag{3.8}$$

$$Pr\left\{(1 - \xi)\hat{p}_t \leq \widetilde{p}_t^{rg} + p_t^{sd} - p_t^{sc}\right\} \geq \beta, \qquad t \in T \tag{3.9}$$

where α, β are the probabilities by which the REG–ESS union output will not violate the upper and lower bounds, respectively. $Pr\{\cdot\}$ stands for the probability that the included inequality constraint meets in consideration of the distribution of the random variable $\widetilde{p}_t^{rg}$. It should be noted that ξ in Eqs. (3.8) and (3.9) has a different meaning from δ in Eq. (3.2); ξ means the confidence interval where the actual output deviation can be satisfied at a given confidence level, whereas δ indicates that the deviation limit should be satisfied without considering the forecast error. Thus ξ can be chosen properly considering the values of α and β.

In Eqs. (3.8) and (3.9), $\widetilde{p}_t^{rg}$ equals the forecast power p_t^{rf} plus the forecast error ε_t. In reference [16], the maximal day-ahead wind forecast error is 30%. We assume that the wind forecast error follows a normal distribution $N(0, \sigma^2)$ [17], where the probability in $[-3\sigma, +3\sigma]$ is 99.73%. The standard deviation σ can be set as $0.1p_t^{wf}$, where p_t^{wf} is the forecasted wind power at time interval t. Meanwhile, wind power output is also limited by the installed capacity of REG (W_{cap}), so the distribution of $\varepsilon_t \in \left[-p_t^{wf}, W_{cap} - p_t^{wf}\right]$ is as illustrated in Figure 3.1.

We consider the more general situation where REG can be wind or solar power. The ε_t follows the conditional probability distribution based on normal distribution, which is given by

$$\Phi_t'(x) = \frac{\Phi_t(x) - \Phi_t\left(-p_t^{rf}\right)}{\Phi_t\left(R_{cap} - p_t^{rf}\right) - \Phi_t\left(-p_t^{rf}\right)} \tag{3.10}$$

where $\Phi_t(x) \sim N\left(0, \left(0.1p_t^{rf}\right)^2\right)$ and R_{cap} is the installed capacity of REG.

Its inverse function is as follows:

$$\Phi_t^{-1'}(x) = \Phi_t^{-1}\left[x\Phi\left(R_{cap} - p_t^{rf}\right) + (1-x)\Phi\left(-p_t^{rf}\right)\right] \tag{3.11}$$

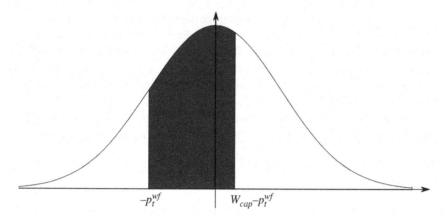

Figure 3.1 Distribution of wind power forecast error.

As functions (3.10) and (3.11) are monotone increasing, constraints (3.8) and (3.9) are equivalent to

$$(1-\xi)\hat{p}_t + \Phi_t'^{-1}(\beta) \le p_t^{sd} - p_t^{sc} + p_t^{rg} \le (1+\xi)\hat{p}_t - \Phi_t'^{-1}(\alpha) \tag{3.12}$$

To ensure that the upper bond is higher than the lower bond, constraints (3.8) and (3.9) should satisfy

$$\Phi_t'^{-1}(\beta) + \Phi_t'^{-1}(\alpha) \le 2\xi\hat{p}_t \tag{3.13}$$

Supposing that $\alpha = \beta$, then the corresponding ξ is shown in Table 3.1.

Another simple way to build chance constraints is as follows:

$$Pr\left\{p_t^{rg} \le \tilde{p}_t^{rg}\right\} \ge \alpha', \quad t \in T \tag{3.14}$$

$$Pr\left\{p_t^{rg} \ge \tilde{p}_t^{rg}\right\} \ge \beta', \quad t \in T \tag{3.15}$$

$$\alpha' + \beta' < 1 \tag{3.16}$$

If the inverse function of the wind forecast error can be found, constraints (3.14) and (3.15) can be transformed into

Table 3.1 Deviation limits under different confidence levels.

$\alpha = \beta$	95%	90%	85%	80%
ξ	39.14%	30.99%	25.24%	20.58%
$\alpha = \beta$	75%	70%	65%	60%
ξ	16.53%	12.87%	9.47%	6.23%

$$F_{\beta'}^{-1}\{\widetilde{p}_t^{rg}\} \leq p_t^{rg} \leq F_{1-\alpha'}^{-1}\{\widetilde{p}_t^{rg}\}, \quad t \in T \qquad (3.17)$$

where $F^{-1}\{\widetilde{p}_t^{rg}\}$ is the inverse function of the forecasted wind power distribution and $F_{\beta'}^{-1}\{\widetilde{p}_t^{rg}\}$ is the quantile corresponding to the confidence β'.

3.3.3 Case Studies on a Union of Wind Farm and Pumped Hydroelectric Storage Plant

3.3.3.1 Simulation Settings

The typical daily power output of a wind farm with an installed capacity of 200 MW is used for the simulation. For simplicity, the electricity prices within a day has two values: the price within the period of 9:00–23:00 is 0.8 RMB kWh^{-1} (RMB is the Chinese currency) and the price within the period of 23:00–9:00 is 0.4 RMB kWh^{-1}. Figure 3.2 illustrates the wind power and electricity price profiles. The parameters of the pumped hydroelectric storage plant (PHSP) are set based on the parameters of the Bath County PHSP in the USA [18] (see Table 3.2).

Six combined pump and turbine generating units (i.e. reversible pump turbines) are installed in Bath County PHS plant and the rated generating power of each unit is about 500 MW. In order to match the installed capacity of the wind farm, the capacity of the PHS plant in the following simulations will be set as 1/24 of the

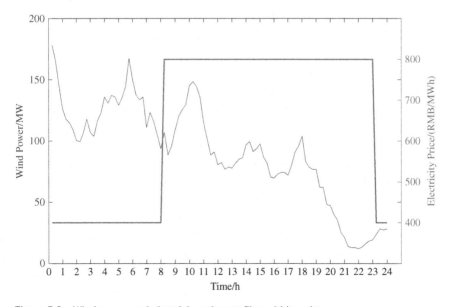

Figure 3.2 Wind power and electricity price profiles within a day.

Table 3.2   Parameters of Bath County PHS plant.

Generating				Pumping			
Water head (m)	Power (MW)	Flow (m³ s⁻¹)	Efficiency	Water head (m)	Power (MW)	Flow (m³ s⁻¹)	Efficiency
Rated: 329.2	380	133.1	90.7%	Rated: 335.3	420	118.9	92.7%
Maximal: 384.0	436	152.9	88.6%	Maximal: 384.0	370	89.5	91.6%
Minimal: 329.2	395	138.8	89.6%	Minimal: 329.2	420	113.3	92.3%
Reservoir							
Maximum level in upper reservoir (m)	Minimum level in upper reservoir (m)		Upper reservoir capacity (Mm³)		Upper adjustment storage (Mm³)		
1012	980		43.8		27.75		
Maximum level in upper reservoir (m)	Minimum level in upper reservoir (m)		Lower reservoir capacity (Mm³)		Lower adjustment storage (Mm³)		
645.57	627.28		37.64		27.75		

capacity of the Bath County PHS plant. It is assumed that there are five units and the rate generating power of each unit is 25 MW. The pumping power of each unit can change within the range 18–22 MW. When the unit is operated as a generator, its power output can be adjusted flexibly and the minimum output is set as 40% of rated power.

The parameters in Table 3.3 indicate that the minimum and maximum volumes ($V^{u,\min}$, $V^{u,\max}$) of the upper reservoir are 16.05 and 43.80 Mm³ and the minimum and maximum volumes ($V^{d,\min}$, $V^{d,\max}$) of the lower reservoir are 9.89 and 37.64 Mm³. Therefore, the dispatchable volume is 27.75 Mm³. The daily change is limited to 1/30 of the dispatchable volume, i.e. 0.93 Mm³.

Existing studies vary a lot on the costs of startup and shutdown for the tubine of the PHS plant. The cost of the PHS turbine per startup is set at about RMB 2000 in

Table 3.3   Parameters of PHS plant in the designed case.

ΔV^u (Mm³)	ΔV^d (Mm³)	$p^{\max}$ (MW)	$p^{\min}$ (MW)	$g^{\max}$ (MW)	$g^{\min}$ (MW)
0.93	0.93	22	18	25	10

some regional power grids in China. Setting the costs for startup and shutdown separately is reasonable because a reversible pump turbine may not complete the whole process of startup and shutdown within a day. Thus, startup and shutdown costs in the following simulations are both RMB 1000 per time.

The penalty factor ω should represent the cost required to compensate for the deviation. According to historical electricity and ancillary service prices of the PJM market, ω is set equal to 0.44 in the simulations.

3.3.3.2 Simulation Results and Comparisons
First, the basic deterministic model is solved and the day-ahead power bidding and the schedules of wind farm and PHS plant operating states can be obtained. The results are shown in Figure 3.3.

It can be seen from this figure that the PHS plant is in charging (pumping) mode from 23:00–8:00 and in discharging (generating) mode from 10:00–11:00, 14:00–17:00, 18:00–20, and 21:00–22:00. Obviously, the dispatch of PHSP is highly related to the electricity prices and takes advantage of price arbitrage, especially during the lower price period. The bidding power fluctuates during the whole day from zero to the maximum power of 200 MW. It is worth noting that the deterministic optimization problem for day-ahead dispatch has multiple optimal

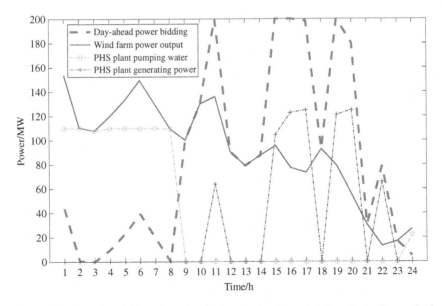

Figure 3.3 Day-ahead dispatch results obtained by the deterministic optimization method.

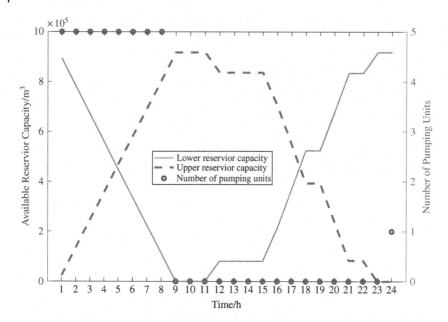

Figure 3.4 Available reservoir capacities and number of pumping units of the PHS plant.

solutions. Without reducing the net revenue, the bidding power profile can be flattened, which can be achieved by a second stage optimization [19].

Figure 3.4 shows the available reservoir capacities and number of pumping units of the PHS plant within the whole day. From 0:00–8:00, all five units are operating in pumping mode and the upper reservoir capacity rises to its maximum. In the high price period, the water in the upper reservoir is gradually released to generate electricity and the available capacity of the lower reservoir recovers correspondingly.

Now, we compare the results obtained by the deterministic optimization model and the two stochastic optimization models. For the chance-constrained optimization, constraints (3.17) are considered and α' and β' are set equal to 0.6 and 0.1, respectively. The scheduled power profiles are given in Figure 3.5.

It can be seen from the figure that the power profiles obtained by deterministic and chance-constrained optimization methods are very close. The power profile obtained using the scenario-based method is obviously different from the profiles by the other two methods, especially during the low electricity price time period. This is mainly because the chance-constraints (3.14) and (3.15) are converted to (3.17) and the whole chance-constrained optimization model is actually a deterministic model. When the upper and the lower bounds of (3.17) and (3.3) are close, then the results of the two models are naturally close.

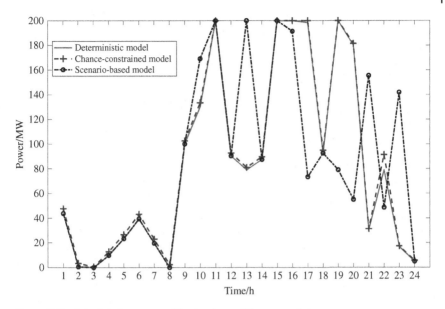

Figure 3.5 Scheduled power profiles by three different models.

Table 3.4 Comparisons of detailed revenues and costs using different models (thousand RMB).

Model	Deterministic model	Chance-constrained model	Scenario-based model
Selling energy revenue	1511	1503	1497
Penalty of deviation	112	101	69
Total startup and shutdown cost	1.2	1.2	1.2
Net revenue	1398	1401	1427

Table 3.4 lists the revenues from selling electricity, penalties of deviation, startup and shutdown costs, and the net revenues corresponding to the three models. The results show that the total startup and shutdown costs of all the models are the same. The revenue of selling electricity and the penalty of deviation using the chance-constrained model are both lower than those of the deterministic model, while net revenue of the former is higher than that of the latter. Although the revenue of selling electricity using the scenario-based model is lower than that

of the other two models, the net revenue of the former is the highest. This is mainly because the influence of the wind power forecast errors is properly considered by a number of scenarios.

3.4 Integrated Bidding Strategies for a REG–ESS Union

3.4.1 Day-Ahead Bidding Strategy

The optimal dispatch model for a REG–ESS union given above only considers the day-ahead schedule. The real-time market prices and the fast response of ESS to the forecast errors on REG power output are not taken into account. In most deregulated electricity markets such as the Scandinavian Nord Pool, market participants usually trade in both day-ahead and balancing markets. In the day-ahead market, power suppliers bid for their generation plan that covers the following day with 12–14 hours prior to the operation day. In real-time operation, the cleared schedules are subject to risk of electricity price variations. In some markets, the participants buy or sell up/down regulation services for any deviation in actual output from the schedule in the balancing market.

Balancing markets can be divided into two categories according to whether the balancing price changes with the imbalance sign [20]. The deviation is traded at a unique price in one-price balancing markets, which is adopted in markets such as the Dutch APX [21]. In the two-price markets such as Nord Pool and the Iberian market, the deviation that is opposite to the system imbalance is traded at the day-ahead price while the imbalance of the same sign with that of the system is settled at the cleared balancing price. We will describe the optimal bidding model based on the second one since it is more comprehensive [22]. The basic idea is illustrated in Figure 3.6.

The inputs of day-ahead bidding include the probabilistic forecasts of REG (wind and/or solar power), day-ahead forecast of electricity prices, and the parameters of the ESS. The REG–ESS union determines the optimal power bidding in the day-ahead market and the ESS charging/discharging dispatch considering the real-time operating conditions.

At the real-time operating stage, the charging/discharging of ESS should take both the REG power output deviation and the electricity prices into account. A reserve-based operating strategy can be employed. The operating state (charging/discharging/idle) of ESS is optimized in advance for each time interval. When imbalance happens, ESS adjusts its state by considering its reserved power and the sign of power imbalance. If the power deviation of REG exceeds the reserved capacity, the ESS can charge or discharge most to the predetermined upper bound, which is often lower than its maximum power capability. The flow

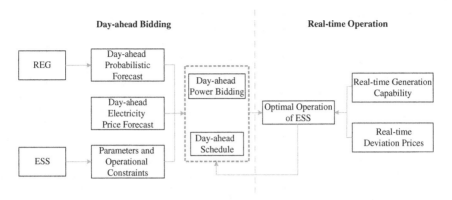

Figure 3.6 Illustration of day-ahead bidding and real-time operation of REG–ESS union.

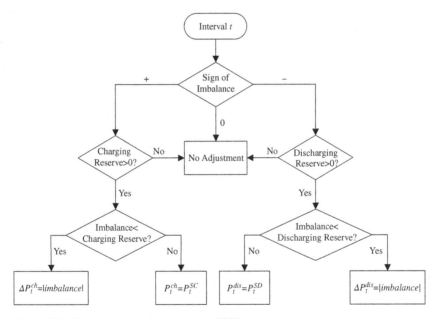

Figure 3.7 Real-time operating strategy of ESS.

chart for the real-time operation of ESS is illustrated in Figure 3.7. Take the decision procedure of charging power adjustment for example. When the REG power at interval t is higher than the bidding value, it is referred to as the positive imbalance of REG (symbolized as "+"), while the opposite situation is called the negative imbalance (symbolized as '−'). In the positive imbalance case, if the ESS has an available charging power reserve from the day-ahead optimization, the charging power adjustment of ESS can be set as the minimum of the imbalance and reserve power. Otherwise, the ESS takes no action. The decision procedure of the discharging power adjustment is similar to that of the charging power adjustment.

Based on the above-mentioned strategies, the optimal bidding problem of the REG-ESS union can be formulated as

$$
\begin{aligned}
\max_{x} \quad & f(x) \\
\text{s.t.} \quad & h(x) \le 0 \\
& g(x) \le 0
\end{aligned}
\tag{3.18}
$$

where $x = \{P^{B}, P^{SC}, P^{SD}\}$ includes all the decision variables, i.e. the vectors of the day-ahead bidding power p_t^b, the charging power p_t^{sc}, and the discharging power p_t^{sd}. The objective function $f(x)$ is the expected revenue of the REG–ESS union, while $h(x) \le 0$ and $g(x) \le 0$ are the considered constraints on the bidding and operation of the REG–ESS union.

The objectives for optimal operation of ESS is to make price arbitrage and to compensate for the potential imbalances of the wind farm. Assuming that the REG output p_t^{rg} at time interval t follows a distribution with probability density function $\varphi_t(p_t^{rg})$, the expected revenue $f(x)$ within a whole day can be formulated as

$$
f(x) = \sum_{t=1}^{T} E\left[S_t(p) \,|\, \lambda_t^D, \lambda_t^{up}, \lambda_t^{dw}, x_t\right] \Delta t
\tag{3.19}
$$

where

$$
\begin{aligned}
&\left[S_t(p) \,|\, \lambda_t^D, \lambda_t^{up}, \lambda_t^{dw}, x_t\right] \\
&= \widehat{\lambda}_t^D P_t^D + \widehat{\lambda}_t^{up} \int_{\underline{rf_t}}^{p_t^b - p_t^{sd}} \left(p - p_t^b + p_t^{sd}\right) \varphi_t(p) dp \\
&\quad + \widehat{\lambda}_t^{dw} \int_{p_t^b + p_t^{sc}}^{\overline{rf_t}} \left(p - p_t^b - p_t^{sc}\right) \varphi_t(p) dp
\end{aligned}
\tag{3.20}
$$

In Eq. (3.19), $S_t(p)$ is the probabilistic revenue for the time interval t, which depends on the day-ahead and regulation prices $\{\lambda_t^D, \lambda_t^{up}, \lambda_t^{dw}\}$, REG output p, day-ahead bidding p_t^b, and ESS power dispatch p_t^{sc}, p_t^{sd}. Assuming that the REG–ESS union is a price taker in the electricity market, the day-ahead energy prices and up/down regulation prices are independent of its bidding decision and power output. Thus, the stochastic prices can be replaced by their expected values. The expected revenue at each time interval t in Eq. (3.20) consists of three parts. The first part is the day-ahead bidding revenue $\widehat{\lambda}_t^D p_t^b$, the second part is the cost of purchasing up-regulation, which is negative in value, and the third part is the revenue of selling down-regulation. Then $\left\{\hat{\lambda}_t^D, \hat{\lambda}_t^{up}, \hat{\lambda}_t^{dw}\right\}$ are the corresponding expected values of $\{\lambda_t^D, \lambda_t^{up}, \lambda_t^{dw}\}$. The parameters with an overhead arc stand for expected values, $\overline{rf}_t$ and $\underline{rf}_t$ are the upper and lower bounds of forecasted REG power at interval t, respectively, while $\varphi_t(\cdot)$ means the probabilistic density function of forecasted REG power.

Now we discuss the corresponding constraints that should be considered. To ensure the validity of the integral lower and upper limits, the following constraints should be satisfied:

$$0 \leq p_t^{sc} \leq u_t^{sc}\left(\overline{wf}_t - p_t^b\right), \quad \forall t \in \{1, 2, ..., T\} \tag{3.21}$$

$$0 \leq p_t^{sd} \leq u_t^{sd}\left(p_t^b - \underline{wf}_t\right), \quad \forall t \in \{1, 2, ..., T\} \tag{3.22}$$

where u_t^{sc} and u_t^{sd} are the binary variables indicating the charging and the discharging states of ESS at time interval t, respectively.

The operational constraints for the ESS should also be included:

$$0 \leq p_t^{sc} \leq u_t^{sc}\overline{p}^{sc}, \quad \forall t \in \{1, 2, ..., T\} \tag{3.23}$$

$$0 \leq p_t^{sd} \leq u_t^{sd}\overline{p}^{sd}, \quad \forall t \in \{1, 2, ..., T\} \tag{3.24}$$

$$u_t^{sc} + u_t^{sd} \leq 1, \quad u_t^{sc}, u_t^{sd} \in \{0, 1\}, \quad \forall t \in \{1, 2, ..., T\} \tag{3.25}$$

$$E_t = E_0 + \sum_{j=1}^{t-1} p_j^{sc} \cdot \eta^{sc} \cdot \Delta t - \sum_{j=1}^{t-1} p_j^{sd}/\eta^{sd} \cdot \Delta t, \quad \forall t \in \{1, 2, ..., T\} \tag{3.26}$$

$$E_{\min} \leq E_t \leq E_{\max}, \quad \forall t \in \{1, 2, ..., T\} \tag{3.27}$$

$$E_T = E_0 \tag{3.28}$$

where (3.23) and (3.24) limit the charging and discharging power at time interval t within the corresponding charging and discharging capacities and $\overline{p}^{sc}$ and $\overline{p}^{sd}$ are the maximum charging and discharging powers of the ESS. Simultaneous charging and discharging are not allowed by (3.25). The energy balance of the ESS is expressed by (3.26) and its residual energy should be within the allowable range $[E_{\min}, E_{\max}]$ at any time with (3.27), while η^{sc} and η^{sc} are the charging and discharging efficiencies of the ESS, respectively. For the daily operation, the residual energy deviation of ESS between the beginning and end of the day will impact on the daily revenue. In Eq. (3.28), the residual energy in ESS is required to be the same at the beginning and the end of the day. This strict constraint could be relaxed to harness more flexibility from the charging and discharging power dispatch of the ESS. Moreover, the ESS degradation effect is not considered here. It should also be noted that the coordination strategies are from the perspective of the REG–ESS union, which is a participant in electricity markets, rather than from the perspective of a system operator. In case network constraints should be considered, the proposed model should take the necessary network constraints into account. When the REG and the ESS are located very near and connected to the same power network bus, constraints limiting their total output below the allowable capacity $\overline{C}_{REG-ESS}$ can be added as

$$p_t^b \leq \overline{C}_{REG-ESS}, \quad \overline{wf}_t - p_t^{sc} \leq \overline{C}_{REG-ESS}$$

If the REG and the ESS are connected to different network buses, similar constraints for the output of REG $\overline{C}_{REG}$ and power of ESS $\overline{C}_{ESS}$ can be added separately when necessary as follows:

$$p_t^b \leq \overline{C}_{REG}, \quad p_t^{sc} \leq \overline{C}_{ESS}, \quad p_t^{sd} \leq \overline{C}_{ESS}$$

3.4.2 Solution Method

As integration is included in the objective function (3.19) and binary variables are included in (3.21) through (3.25), the formulation is a mixed integer nonlinear model, which cannot be solved directly. A scenario-based method can be used to convert (3.20) into a linear equation and the whole problem will be a mixed integer linear program, which can be solved by off-the-shelf software like CPLEX [23].

In reference [22], a modified gradient descent algorithm is employed to solve the problem according to the characteristics of the formulation. It should be noted that, as the problem may be non-convex, in that case, the algorithm may only obtain the nearest local optimal solution rather than the global optimal solution. However, the computational effort required for the algorithm is considerably less than metaheuristic algorithms, such as the genetic algorithm, which can only theoretically obtain global optimal solutions. As the gradient descent algorithm is computationally friendly, a number of initial solutions can be tried to obtain a better solution. The procedure of this algorithm is illustrated as follows.

Algorithm

Modified Gradient Descent Algorithm
```
Begin:
    {P^B0, P^SD0, P^SC0} = {P^Bset, 0, 0}
    S^0 = f(P^B0, P^SD0, P^SC0)
    for k = 1 : K
Execute ESS charging/discharging optimization
⇒{ [ΔP^SDk, ΔP^SCk] } (see Section 3.4.2.1)
Execute Bidding power optimization ⇒{ΔP^Bk}
(see Section 3.4.2.2)
```

$$\mathbf{P}^{Bk} = \mathbf{P}^{B(k-1)} + \Delta\mathbf{P}^{Bk}$$
$$\mathbf{P}^{SDk} = \mathbf{P}^{SD(k-1)} + \Delta\mathbf{P}^{SDk}$$
$$\mathbf{P}^{SCk} = \mathbf{P}^{SC(k-1)} + \Delta\mathbf{P}^{SCk}$$
$$S^k = f\left(\mathbf{P}^{Bk}, \mathbf{P}^{SDk}, \mathbf{P}^{SCk}\right)$$

```
if S^k - S^(k-1) < 0
    Execute back-tracking procedure:
```
$$\gamma = 1, \quad S^k(1) = S^k$$
```
        while
```
$$S^k(\gamma) \le S^{k-1}$$
$$\gamma = \beta\gamma$$
$$\mathbf{P}^{SDk}(\gamma) = \mathbf{P}^{SD(k-1)} + \left(1 - (1-\gamma)\mathbf{u}^{SDk}\right)\Delta\mathbf{P}^{SDk}$$
$$\mathbf{P}^{SCk}(\gamma) = \mathbf{P}^{SC(k-1)} + \left(1 - (1-\gamma)\mathbf{u}^{SCk}\right)\Delta\mathbf{P}^{SCk}$$
$$\mathbf{P}^{Bk}(\gamma) = \mathbf{P}^{B(k-1)} + \gamma\Delta\mathbf{P}^{Bk}$$
$$S^k(\gamma) = f\left(\mathbf{P}^{Bk}(\gamma), \mathbf{P}^{SDk}(\gamma), \mathbf{P}^{SCk}(\gamma)\right)$$
```
        end
```
$$\mathbf{P}^{SDk} = \mathbf{P}^{SDk}(\gamma), \quad \mathbf{P}^{SCk} = \mathbf{P}^{SCk}(\gamma)$$
$$\mathbf{P}^{Bk} = \mathbf{P}^{Bk}(\gamma), \quad S^k = S^k(\gamma)$$
```
    else if
```
$$0 \le S^k - S^{(k-1)} \le \varepsilon$$
```
        stop
    else
        continue the loop
    end
end
```
$$\mathbf{P}^{SD} = \mathbf{P}^{SDk}, \quad \mathbf{P}^{SC} = \mathbf{P}^{SCk}, \quad \mathbf{P}^{B} = \mathbf{P}^{Bk}$$
```
end
```

The initial bidding $\mathbf{P}^{Bset}$ can be set as the expected forecast power of REG. In reference [24], $\mathbf{P}^{Bset}$ is chosen as the optimal quantile from wind power predictive densities. The parameter γ is the step size, whose initial value is one. The back-tracking factor β shortens the step size. The key point is that the modified back-tracking will not change the value of u_t^{sc} and u_t^{sd}. Take u_t^{sc} for example. If $u_t^{sc}(1) = 0$, then $p_t^{sck}(\gamma) = p_t^{sck}(1) = 0$ and $u_t^{sc}(\gamma) = 0$. Otherwise, $p_t^{sck}(\gamma) = p_t^{sc(k-1)} + \gamma\Delta p_t^{sck} > 0$ and $u_t^{sc}(\gamma) > 0$.

In Eq. (3.20), when the expected prices and REG power distributions are available through predictions, then $E\left[S_t(p) \mid \lambda_t^D, \lambda_t^{up}, \lambda_t^{dw}, x\right]$ is a function of p_t^b, p_t^{sd}, and p_t^{sc}, denoted by $S(p_t^b, p_t^{sd}, p_t^{sc})$. At any time interval t, the expected revenue $S(p_t^b, p_t^{sd}, p_t^{sc})$ can be linearly approximated as

$$S\left(p_t^b, p_t^{sd}, p_t^{sc}\right) \approx S\left(p_t^{b0}, p_t^{sd0}, p_t^{sc0}\right)$$
$$+ \left[\frac{\partial S}{\partial p_t^{b0}}, \frac{\partial S}{\partial p_t^{sd0}}, \frac{\partial S}{\partial p_t^{sc0}}\right] \cdot \left[\Delta p_t^b, \Delta p_t^{sd}, \Delta p_t^{sc}\right]^T \tag{3.29}$$

and

$$\frac{\partial S_t}{\partial p_t^b} = \overline{\lambda}_t^D - \overline{\lambda}_t^{up} \Gamma_t \left(p_t^b - p_t^{sd}\right) - \overline{\lambda}_t^{dw} \left[1 - \Gamma_t \left(p_t^b + p_t^{sc}\right)\right]$$

$$\frac{\partial S_t}{\partial p_t^{sc}} = \overline{\lambda}_t^{up} \Gamma_t \left(p_t^b - p_t^{sd}\right)$$

$$\frac{\partial S_t}{\partial p_t^{sd}} = -\overline{\lambda}_t^{dw} \left[1 - \Gamma_t \left(p_t^b + p_t^{sc}\right)\right]$$

The first part of (3.29) is fixed, so the optimal solution of this function can be obtained by solving the second part. $\Gamma_t(\cdot)$ is the cumulative distribution function of the forecasted REG power at time interval t. Moreover, Eqs. (3.21) and (3.22) are quadratic, which can be converted to linear equations via an iterative procedure on the corresponding decision variables. For the ESS charging/discharging optimization, p_t^b will be fixed at the value obtained from the previous iteration, while for the bidding power optimization, the values of u_t^{sc}, u_t^{sd} are considered to be known parameters.

3.4.2.1 ESS Charging/Discharging Optimization

In this part of the optimization, the decision variables all regard charging and discharging of the ESS ($\Delta p_t^{sdk}, \Delta p_t^{sck}$). The original optimization problem (3.19)–(3.28) can be linearized as follows:

$$\max \sum_{t=1}^{T} \left(\frac{\partial S_t}{\partial p_t^{sd(k-1)}} \Delta p_t^{sdk} + \frac{\partial S_t}{\partial p_t^{sc(k-1)}} \Delta p_t^{sck}\right) \tag{3.30}$$

$$\text{s.t.} \, 0 \leq p_t^{sc(k-1)} + \Delta p_t^{sck} \leq u_t^{sck} \overline{p}_t^{sc} \tag{3.31}$$

$$0 \leq p_t^{sd(k-1)} + \Delta p_t^{sdk} \leq u_t^{sdk} \overline{p}_t^{sd} \tag{3.32}$$

$$0 \leq p_t^{sc(k-1)} + \Delta p_t^{sCk} \leq u_t^{sck} \left(\overline{wf}_t - p_t^{b(k-1)}\right) \tag{3.33}$$

$$0 \leq p_t^{sd(k-1)} + \Delta p_t^{sdk} \leq u_t^{sdk} \left(p_t^{b(k-1)} - \underline{wf}_t\right) \tag{3.34}$$

$$u_t^{sck} + u_t^{sdk} \leq 1 \tag{3.35}$$

$$E_t^k = E_t^{k-1} + \sum_{j=1}^{t-1} \Delta p_t^{sck} \cdot \eta^{sc} \cdot \Delta t - \sum_{j=1}^{t-1} \Delta p_t^{sdk} / \eta^{sd} \cdot \Delta t \tag{3.36}$$

$$E_{\min} \leq E_t^k \leq E_{\max} \tag{3.37}$$

$$E_T^k = E_0 \tag{3.38}$$

where the superscript k stands for the iteration number k, where all the parameters with superscript $(k - 1)$ are constants. It should be noted that the obtained optimal bidding $\left\{ p_t^{b(k-1)} \right\}$ at iteration $(k - 1)$ is used in Eqs. (3.33) and (3.34), which makes the constraints linear.

3.4.2.2 Bidding Power Optimization

In this part of the optimization, the decision variables are the bidding-related variables, i.e. Δp_t^{bk}. The objective function only contains the revenue from the day-ahead market and the optimization problem is formulated as follows:

$$\max \sum_{t=1}^{T} \frac{\partial S_t}{\partial p_t^{b(k-1)}} \Delta p_t^{bk} \tag{3.39}$$

$$\text{s.t.} \, p_t^{b(k-1)} + \Delta p_t^{bk} \leq \overline{wf}_t - p_t^{sc(k-1)} - \Delta p_t^{sck} \tag{3.40}$$

$$p_t^{sd(k-1)} + \Delta p_t^{sdk} + \underline{wf}_t \leq p_t^{b(k-1)} + \Delta p_t^{bk} \tag{3.41}$$

The change of bidding power is limited by the power variation range of the REG–ESS union. The latest values of $\Delta p_t^{sdk}, \Delta p_t^{sck}$ are considered in the above formulation.

3.4.3 Illustrative Example

In this small example, only three time intervals are considered for the sake of simplicity and transparency. A wind farm and a battery energy storage (WF–BESS) union is considered. The basic data are given in Tables 3.5 and 3.6.

It is assumed that potential wind power generation obeys a uniform distribution. Because there is no limit on the distribution in Equation (3.29), other distributions can also be adapted to the proposed method. The assumption of uniformity is not very grounded, but it is the simplest, especially when integrated into objective functions. The main purpose of employing a uniform distribution here is to compare the optimality of the algorithm discussed above with commercial software

Table 3.5 Parameters of the WF-BESS union.

η^{sc}	η^{sd}	E_{min} [MWh]	E_{max} [MWh]
0.9	0.9	1	10
E_0 [MWh]	$\overline{p}^{sc}$ [MW]	$\overline{p}^{sd}$ [MW]	β
5	10	10	0.1

Table 3.6 Parameters of market prices and wind power ranges.

Time interval	$\bar{\lambda}^D$ [/MWh]	$\bar{\lambda}^{dw}$ [/MWh]	$\bar{\lambda}^{up}$ [/MWh]	$\overline{Wf}$ [MW]	$\underline{Wf}$ [MW]
1	57.1	28.5	71.4	90	0
2	114.2	100	14.4	60	0
3	85.7	71.4	100	75	0

such as CPLEX. Substituting $p \rightarrow U\left(wf_t, \overline{wf}_t\right)$ into Eq. (3.20), the objective function is

$$S_t = \bar{\lambda}_t^D p_t^b - \frac{\bar{\lambda}_t^{up}}{2\overline{wf}_t}\left(p_t^b - p_t^{sd}\right)^2 + \frac{\bar{\lambda}_t^{dw}}{2\overline{wf}_t}\left(\overline{wf}_t - p_t^b - p_t^{sc}\right)^2 \tag{3.42}$$

Together with constraints (3.21) through (3.28), a quadratic formulation can be built and solved by CPLEX.

The optimal solutions are listed in Table 3.7. It can be seen from this table that with the coordination of ESS and WF, the union tends to bid a higher amount in the day-ahead market. As the price in interval 2 is higher, the ESS sets some discharging reserve in this period and charging reserve in intervals 1 and 3. The increasing amount of bidding in interval 2 is also considerably higher than that of intervals 1 and 3.

As the latest information $\left\{\Delta p_t^{sck}, \Delta p_t^{sdk}\right\}$ of charging and discharging reserve is utilized in the bidding model, the converging speed of the proposed algorithm is considerably faster than the conventional gradient descent algorithm, which only uses the information of $p_t^{sc(k-1)}, p_t^{sd(k-1)}$ in iteration k (i.e. replace $\Delta p_t^{sck}, \Delta p_t^{sdk}$ by $\Delta p_t^{sc(k-1)}, \Delta p_t^{sd(k-1)}$ in Eqs. (3.40) and (3.41)). As shown in Figure 3.8, it takes the modified gradient descent algorithm about six iterations to converge, and the objective value is 8630$.

Table 3.7 Comparison of the bidding results.

Interval	1	2	3
Bidding without BESS (MW)	60.0	20.0	37.5
Bidding with BESS (MW)	63.7	47.0	48.5
Charging power of BESS (MW)	5.6	0	4.4
Discharging power of BESS (MW)	0	8.1	0

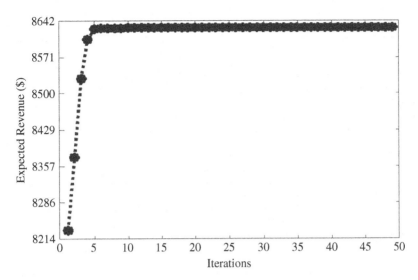

Figure 3.8 Converging process of the modified gradient descent algorithm.

The Hessian matrix of the objective function (3.42) is

$$
H_t = \frac{1}{\overline{\overline{wf}}}
\begin{bmatrix}
-\overline{\lambda}_t^{up} + \overline{\lambda}_t^{dw} & \overline{\lambda}_t^{up} & \overline{\lambda}_t^{dw} \\
-\overline{\lambda}_t^{up} & -\overline{\lambda}_t^{up} & 0 \\
\overline{\lambda}_t^{dw} & 0 & \overline{\lambda}_t^{dw}
\end{bmatrix}
$$

As the third-order leading principal minor is positive, the Hessian matrix is not negative definite, so the problem is nonconvex. CPLEX only gives the upper bound, which is 8657\$. Meanwhile, the solution obtained with our proposed algorithm is 8630\$, which is quite close to that upper bound.

3.5 Conclusion and Discussion

In this chapter, the optimal day-ahead bidding strategies are discussed for the union of REG and ESS. First, a basic deterministic model is given considering the coordination of REG and ESS to maximize their revenue. Then the scenario-based and chance-constrained stochastic optimization models are derived to consider uncertainties and forecast errors of renewable power generation. These models only consider the day-ahead schedule, while the change of electricity prices and the adjustment of ESS's power output in real time are not taken into account. A comprehensive bidding strategy is then introduced, which considers the two-price balancing market rule. A mixed integer nonlinear optimization formulation is built to determine optimal offers considering expected wind power forecasting errors and the power balancing capability of the ESS. A modified gradient descent algorithm is designed to solve this nonlinear programming problem. A simple example is given to illustrate the effectiveness of the method for getting the optimal bidding solution.

References

1 Perez, D.J.I., Perea, A., and Wilhelmi, J.R. (2010). Optimal short-term operation and sizing of pumped-storage power plants in systems with high penetration of wind energy. *2010 7th International Conference on the European Energy Market*, Madrid (23–25 June 2010).

2 Garcia-Gonzalez, J., de la Muela, R.M.R., Santos, L.M. et al. (2008). Stochastic joint optimization of wind generation and pumped-storage units in an electricity market. *IEEE Transactions on Power Systems* 23 (2): 460–468.

3 Castronuovo, E.D. and Lopes, J.A.P. (2004). Optimal operation and hydro storage sizing of a wind-hydro power plant. *International Journal of Electrical Power and Energy Systems* 26 (10): 771–778.

4 Tuohy, A. and O'Malley, M. (2009). Impact of pumped storage on power systems with increasing wind penetration. *IEEE Power and Energy Society General Meeting*, Calgary (26–30 July 2009).

5 Contaxis, G. and Vlachos, A. (2000). Optimal power flow considering operation of wind parks and pump storage hydro units under large scale integration of renewable energy sources. *IEEE Power Engineering Society Winter Meeting*, Singapore (23–27 January 2000).

6 Castronuovo, E.D., Usaola, J., Bessa, R. et al. (2013). An integrated approach for optimal coordination of wind power and hydro pumping storage. *Wind Energy* 17 (6): 829–852.

7 Pinson, P., Papaefthymiou, G., Klöckl, B., and Verboomen, J. (2009). Dynamic sizing of energy storage for hedging wind power forecast uncertainty. *2009 IEEE Power and Energy Society General Meeting, PES '09*. IEEE, pp. 1–8.

8 Bitar, E., Rajagopal, R., Khargonekar, P., and Poolla, K. (2011).The role of co-located storage for wind power producers in conventional electricity markets. *Proceedings of the 2011 American Control Conference*, pp. 3886–3891.

9 Bathurst, G.N. and Strbac, G. (2003). Value of combining energy storage and wind in short-term energy and balancing markets. *Electric Power Systems Research* 67 (1): 1–8.

10 Foley, A.M., Leahy, P.G., Marvuglia, A. et al. (2012). Current methods and advances in forecasting of wind power generation. *Renewable Energy* 37 (1): 1–8.

11 Antonanzas, J., Osorio, N., Escobar, R. et al. (2016). Review of photovoltaic power forecasting. *Solar Energy* 136: 78–111.

12 Wang, Z., Shen, C., and Liu, F. (2018). A conditional model of wind power forecast errors and its application in scenario generation. *Applied Energy* 212: 771–785.

13 Zhang, S., Song, Y., and Hu, Z. (2011). Robust optimization method based on scenario analysis for unit commitment considering wind uncertainties. *2011 IEEE Power and Energy Society General Meeting*, Michigan (24–29 July 2011).

14 Sharma, K.C., Jain, P., and Bhakar, R. (2013). Wind power scenario generation and reduction in stochastic programming framework. *Electric Power Components & Systems* 41 (3): 271–285.

15 Charnes, A. and Cooper, W.W. (1959). Chance-constrained programming. *Management Science* 6: 73–79.

16 Korpas, M. and Holen, A.T. (2006). Operation planning of hydrogen storage connected to wind power operating in a power market. *IEEE Transactions on Energy Conversion* 21 (3): 742–749.

17 Feng, G., Hallam, A., and Chien-Ning, Y. (2009). Wind generation scheduling with pump storage unit by Collocation method. *IEEE Power and Energy Society General Meeting*, Calgary (26–30 July 2009).

18 Fostiak, R.J. and Thompson, W.L. (1982). Electrical features of the Bath County pumped-storage project. *IEEE Transactions on Power Apparatus and Systems* PAS-101 (9): 3166–3172.

19 Luo, Z., Hu, Z., Song, Y. et al. (November 2013). Optimal coordination of plug-in electric vehicles in power grids with cost-benefit analysis—part I: enabling techniques. *IEEE Transactions on Power Systems* 28 (4): 3546–3555.

20 Morales, J.M., Conejo, A.J., Madsen, H. et al. (2014). *Integrating Renewables in Electricity Markets – Operational Problems*. Springer.

21 Zugno, M., Morales, J.M., Pinson, P., and Madsen, H. (2013). Pool strategy of a price-maker wind power producer. *IEEE Transactions on Power Apparatus and Systems* 28 (3): 3440–3450.

22 Ding, H., Pinson, P., Hu, Z., and Song, Y. (January 2016). Integrated bidding and operating strategies for wind-storage systems. *IEEE Transactions on Sustainable Energy* 7 (1): 163–172.

23 IBM ILOG CPLEX Optimization Studio. https://www.ibm.com/products/ilog-cplex-optimization-studio.

24 Pinson, P., Chevallier, C., and Kariniotakis, G.N. (2007). Trading wind generation from short-term probabilistic forecasts of wind power. *IEEE Transactions on Power Apparatus and Systems* 22 (3): 1148–1156.

4

Refined Bidding and Operating Strategy for a Renewable Energy Generation and Energy Storage System Union

4.1 Introduction

In last chapter, it was shown that the energy storage system (ESS) can be coordinated with renewable energy generation (REG) in a day-ahead electricity market to increase the total revenue of the REG–ESS union. Some published research works assumed that the role of ESS in the union was to compensate for deviations from a predefined operation schedule and to smooth power output in a real-time operation [1, 2]. They overlooked the opportunity of optimizing market participation by optimally foreseeing how ESS could jointly accommodate deviations from schedule and perform arbitrage. To prevent ESS from charging at high-price periods and discharging at low-price periods when balancing deviations, a strategy was introduced in Chapter 3 that embeds the balancing strategy into day-ahead optimization and can effectively increase the revenue of the REG–ESS union. These methods focus more on REG power deviations, while they are not very sensitive to balancing prices. In contrast, if aiming to integrate complicated real-time control strategies into day-ahead offering/bidding optimization, methods such as stochastic dynamic programming may suffer from very heavy computing efforts. Linear decision rules permit decisions to be defined through affine transformation of realizations of uncertain parameters [3], hence keeping computational costs low. In this chapter, linear decision rules as real-time control strategies of the REG–ESS union will be introduced [4], where past, current, and updated forecast information can be fully utilized and linearly combined to obtain optimal operation policies. This information includes deviations in day-ahead and balancing prices, as well as wind power generation, from their originally predicted values. Adopting linear decision rules as the real-time operation strategy guarantees tractability while accommodating the dynamic nature of the operational problem.

The risk in electricity markets for REG can be reduced by coordinating with ESS. Meanwhile, wind power forecast errors generally increase with the extension of

Energy Storage for Power System Planning and Operation, First Edition. Zechun Hu.

the forecast time horizon, which means that the forecast results become more accurate when it comes closer to the real-time. Consequently, except for coordinating REG with ESS in the day-ahead market, rolling dispatch/operation can also be utilized. Reference [5] proposes a rolling unit commitment methodology and examines potential cost savings by committing units more frequently. In some market designs, REGs can revise their offerings to mitigate the deviation penalty brought by the day-ahead forecast errors, and the system operator (SO) can reduce reserve capacity for the system [6, 7]. Therefore, intraday reoffering can be beneficial to both REGs and system operation [8].

Although there are quite a lot of researches studying these two aspects separately, few of them have considered the optimal operation of REG and ESS in a market that allows for rolling energy offering/bidding. In fact, the Spanish power market as well as the Nordic one has an intraday energy rebidding section. It is organized six times a day and biddings can be submitted about four hours in advance, covering the upcoming 24 hours [9]. Forecast error of wind and solar power increases distinctly with the increasing span of time horizon, so the intraday forecast power of a longer horizon is no more credible than the corresponding day-ahead value. Besides, the intraday market opens six times a day. REGs should revise their output plan repeatedly, but only the biddings for the near future (for example, four to eight hours) will become effective. In the second part of this chapter, a modified market design will be introduced to allow for both day-ahead and intraday biddings. Then coordinated operation of REG and ESS will be discussed in detail. Rolling stochastic optimization formulations for day-ahead, intraday offerings, and real-time operations are built to obtain the optimal offering strategies of an REG–ESS union in all bidding sections in an attempt to maximize its overall revenue.

4.2 Real-Time Operation with Linear Decision Rules

The whole time horizon (one day, for example) can be discretized into T time intervals, usually with one hour, half an hour, or fifteen minutes for each interval and corresponding to the market time units. The state vector $x = [x_1, ..., x_T]^{\mathrm{T}} \in \mathbb{R}^T$ stands for the residual energy of ESS at the end of each time interval. The state variables are temporally coupled as:

$$x_t = x_{t-1} + Bu_t, \qquad t = 1, 2, ..., T \tag{4.1}$$

where $B = [0, \eta_c, -1/\eta_d]$ and x_0 is the initial residual energy; $u_t = [p_t^{rg}, p_t^{sc}, p_t^{sd}]^{\mathrm{T}}$ consists of the tth elements of REG's power output vector p^{rg}, charging power vector p^{sc}, and discharging power vector p^{sd}, where $p^{rg} = [p_1^{rg}, ..., p_T^{rg}]^{\mathrm{T}}$,

$\boldsymbol{p}^{sc} = \left[p_1^{sc}, ..., p_T^{sc}\right]^{\mathrm{T}}$, and $\boldsymbol{p}^{sd} = \left[p_1^{sd}, ..., p_T^{sd}\right]^{\mathrm{T}}$; and η_c and η_d are the charging and discharging efficiencies of the ESS.

According to linear decision rules [3], the power vectors of an REG–ESS union are determined by the affine function of day-ahead and balancing price forecast errors, as well as wind power forecast error, i.e.

$$\begin{bmatrix} \boldsymbol{p}^{rg} \\ \boldsymbol{p}^{sc} \\ \boldsymbol{p}^{sd} \end{bmatrix} = \begin{bmatrix} \hat{\boldsymbol{p}}^{rg} \\ \hat{\boldsymbol{p}}^{sc} \\ \hat{\boldsymbol{p}}^{sd} \end{bmatrix} + \begin{bmatrix} \boldsymbol{D}_{da}^{rg} & \boldsymbol{D}_{rt}^{rg} & \boldsymbol{D}_{rf}^{rg} \\ \boldsymbol{D}_{da}^{sc} & \boldsymbol{D}_{rt}^{sc} & \boldsymbol{D}_{rf}^{sc} \\ \boldsymbol{D}_{da}^{sd} & \boldsymbol{D}_{rt}^{sd} & \boldsymbol{D}_{rf}^{sd} \end{bmatrix} \begin{bmatrix} \Delta\boldsymbol{\pi}^{da} \\ \Delta\boldsymbol{\pi}^{rt} \\ \Delta\boldsymbol{p}^{rg} \end{bmatrix} \qquad (4.2)$$

and can be denoted as $\boldsymbol{p} = \hat{\boldsymbol{p}} + \boldsymbol{D}\boldsymbol{\delta}$, where $\hat{\boldsymbol{p}}$ is the nominal power vector and $\boldsymbol{\delta}$ is the vector consisting of forecast errors for day-ahead prices $\Delta\boldsymbol{\pi}^{da}$, balancing prices $\Delta\boldsymbol{\pi}^{rt}$, and REG power $\Delta\boldsymbol{p}^{rg}$. The affine matrix $\boldsymbol{D}$ consists of nine submatrices, which link power adjustments with forecast errors of wind power and prices. Take $\boldsymbol{D}_{rt}^{rg}$, for example, which represents the influence of the balancing price forecast error on wind power adjustment. We also denote $\boldsymbol{D}^{rg} = \left[\boldsymbol{D}_{da}^{rg}, \boldsymbol{D}_{rt}^{rg}, \boldsymbol{D}_{rf}^{rg}\right]$, $\boldsymbol{D}^{sc} = \left[\boldsymbol{D}_{da}^{sc}, \boldsymbol{D}_{rt}^{sc}, \boldsymbol{D}_{wf}^{sc}\right]$, and $\boldsymbol{D}^{sd} = \left[\boldsymbol{D}_{da}^{sd}, \boldsymbol{D}_{rt}^{sd}, \boldsymbol{D}_{rf}^{sd}\right]$, which are submatrices making up the matrix $\boldsymbol{D}$ for later use. In the day-ahead optimization, as the forecast of wind power and prices will be formulated by scenarios, the forecast errors in Eq. (4.2) are the differences between the correspondingly values of each scenario and their mean values.

As shown in Figure 4.1, linear decision rules are adopted for both day-ahead optimization and real-time operation. Denoted by $\tilde{\boldsymbol{\pi}}^{da}$, $\tilde{\boldsymbol{\pi}}^{rt}$, and $\tilde{\boldsymbol{\pi}}^{rg}$, the forecast vectors of day-ahead prices, balancing prices, and wind power are gathered 12–36 hours before the dispatch. These data are fed to the optimization model that will be introduced to obtain the optimal offering, $\boldsymbol{p}^{bid}$, as well as the parameters of the linear decision rules, $\hat{\boldsymbol{p}}$ and $\boldsymbol{D}$. Then $\hat{\boldsymbol{p}}$ and $\boldsymbol{D}$ are applied for the real-time operation. At time interval t, realizations of day-ahead prices, $\boldsymbol{\pi}^{da}$, some balancing prices, and REG power of the operating day, $\boldsymbol{\pi}^{rt}, \boldsymbol{p}^{rg}$, are known, while the intraday forecasts of balancing prices and REG power, $\hat{\boldsymbol{\pi}}^{rt}, \hat{\boldsymbol{p}}^{rg}$, are updated. These data together with the day-ahead forecasts are used to determine the power output of REG and ESS during real-time operation. Take the REG power in time interval t for example, its decision is obtained by the following equation:

$$p_t^{rg} = \hat{p}_t^{rg} + \sum_{i=1}^{T} D_{da,t,i}^{rg}\left(\pi_i^{da} - \mathbb{E}\left(\tilde{\pi}_i^{da}\right)\right) + \sum_{i=1}^{t} D_{rt,t,i}^{rg}\left(\pi_i^{rt} - \mathbb{E}\left(\tilde{\pi}_i^{rt}\right)\right) +$$

$$\sum_{j=t+1}^{T} D_{rt,t,j}^{rg}\Delta\mathbb{E}\left(\tilde{\pi}_j^{rt}\right) + \sum_{i=1}^{t} D_{rf,t,i}^{rg}\left(p_i^{rg} - \mathbb{E}\left(\hat{p}_i^{rf}\right)\right) + \sum_{j=t+1}^{T} D_{rf,t,j}^{rg}\Delta\mathbb{E}\left(\tilde{\pi}_j^{rf}\right)$$

$$(4.3)$$

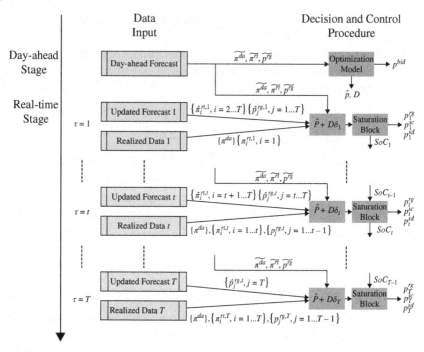

Figure 4.1 Illustration of the decision procedure and control of an REG–ESS union based on linear decision rules.

where $D^{rg}_{da,t,i}$ is the element at row t, column i in matrix $\boldsymbol{D}^{rg}_{da}$. In addition,

$$\Delta\mathbb{E}\left(\widetilde{\pi}^{rt}_j\right) = \mathbb{E}\left(\hat{\pi}^{rt}_j\right) - \mathbb{E}\left(\widetilde{\pi}^{rt}_j\right)$$

$$\Delta\mathbb{E}\left(\widetilde{p}^{rg}_j\right) = \mathbb{E}\left(\hat{p}^{rg}_j\right) - \mathbb{E}\left(\widetilde{p}^{rg}_j\right)$$

Unlike the notation in Figure 4.1, the superscripts indicating the occurrence of an update for $\left\{\hat{\pi}^{rt}_j\right\}$ and $\left\{\hat{\pi}^{rt}_j\right\}$ are omitted in Eq. (4.3) for the sake of succinctness. The charging or discharging decisions for ESS are made in a similar manner, simply replacing the superscript "*rg*" of p and D in the above equations with "*sc*" or "*sd*".

The decision about power output at any time interval consists of four parts, as shown in Eq. (4.3). The first part is the nominal power $\hat{p}^{rg}_t$, determined from the day-ahead offering stage. The second part corresponds to the influence of the day-ahead price forecast error, while the last two parts (corresponding to the third plus the fourth items and the fifth plus the sixth items, respectively) are for the influence of the forecast errors in the balancing prices and real-time available REG power. When it comes to real-time, the day-ahead prices $\{\pi_i^{da}\}$ for the whole time horizon are known. However, the balancing prices $\{\pi_i^{rt}\}$ and the available REG power $\{p_i^{rg}\}$ are only accessible for the past hours, $\{i = 1, ..., t\}$. Consequently, those for the following intervals, $\{i = t + 1, ..., T\}$, should be substituted with the expectation of an updated forecast $\{\mathbb{E}(\hat{\pi}_i^{rt})\}$ and $\{\mathbb{E}(\hat{p}_i^{rg})\}$ to obtain the deviations. It should be noted that for each time interval, the latest updated $\hat{\pi}_i^{rt}$ and $\hat{p}_i^{rg}$ are used.

Stochastic parameters are commonly dealt with by constraining their potential realizations within specific uncertainty sets, around their expectations [10]. Although the polyhedron is a good choice to describe uncertainty sets, if the given prices and wind power forecasting error were too large, the only feasible matrix $\boldsymbol{D}$ would be an all-zero matrix. Then the strategy could not react to the real-time information, and the strategy on linear decision rules would lose its value. In order to prevent this from happening, the formulation of uncertainty sets is tightened here, which may result in the decision vector violating the operation of all constraints of the REG–ESS union during operation. The violation comes from the fact that $\boldsymbol{D}$ and $\hat{\boldsymbol{p}}$ are predetermined, while $\boldsymbol{\delta}$ is a random vector, which cannot be completely covered by scenarios or perfectly formulated by polyhedral constraints due to the above-mentioned reasons. Consequently, an examination through a saturation block is necessary to constrain the charging and discharging power of ESS. As illustrated in Figure 4.1, the charging and discharging power of ESS obtained through linear decision rules, as well as the state of charge (SoC), are fed to the saturation block in order to prevent the ESS from over-charging or over-discharging. At any time, the charging and discharging power should be limited so as not to drive the residual energy out of the allowable ranges during the next time interval. Consequently, the charging threshold is

$$\min\left\{\overline{P}_c, \frac{E_{\max} - E_t}{\eta_c \Delta t}\right\}$$

and the discharging threshold is

$$\min\left\{\overline{P}_d, \frac{E_t - E_{\min}}{\Delta t}\eta_d\right\}$$

As the saturation block restricts the power output only in extreme situations, which occurs rarely, it does not have much influence on the overall revenue of the REG–ESS union. Consequently, the saturation block is not included in the optimization model.

4.3 Optimal Offering Strategy with Linear Decision Rules

An REG–ESS union makes its day-ahead bidding decision based on the forecast REG power profile and electricity prices. In the balancing markets, the union will dynamically adjust the power outputs of REG and ESS based on the linear decision rules, where the known forecast errors of the REG power profile and electricity prices are used. In order to maximize the total revenue in day-ahead and real-time stages, the real-time control strategy should be taken into account in the day-ahead bidding decision, where the optimal control parameters are obtained. As illustrated in Figure 4.2, the day-ahead forecast, day-ahead optimization, and real-time operation are considered as a whole.

When determining the optimal day-ahead offering, the REG–ESS union should not only pursue higher expected revenue but also reduce the potential risk. In the following, the risk is formulated by a conditional value at risk (CVaR) [11], which is co-optimized with the expected revenue through linear combination. The optimization problem can be formulated as

$$\max_{\theta} \quad \gamma \mathbb{E}\left(\widetilde{S}(\theta,\delta)\right) + (1-\gamma)\text{CVaR}_{\alpha}\left(\widetilde{S}(\theta,\delta)\right) \qquad (4.4)$$

$$\text{s.t.} \quad g(\theta,\delta) \leq 0 \qquad (4.5)$$

where $\theta = \{p^{bid}, D, \hat{p}\}$ are decision variables and p^{bid} is the vector of day-ahead offers p_t^{bid}. Here, it should be noted that as the REG–ESS union adopts the price-taker strategy, it will bid at zero or slightly negative prices to guarantee its bidding to be cleared and to get paid at the cleared prices. Consequently, it only needs to determine the optimal quantity to offer. The objective function considers the expectation and risk of the REG–ESS revenue, while (4.5) limits the power of REG and ESS, the residual energy of ESS, and offering quantities.

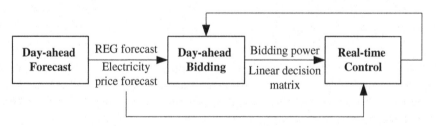

Figure 4.2 Illustration of day-ahead decision considering real-time control strategy.

4.3.1 Objective Function

The bidding objective of the REG–ESS union is to maximize its total revenue from both day-ahead and balancing stages. Risk is another important issue to consider. $\mathbb{E}\left(\widetilde{S}(\boldsymbol{\theta},\boldsymbol{\delta})\right)$ denotes the expected revenue driven by the uncertainty of $\boldsymbol{\delta}$, while $\mathrm{CVaR}_{\alpha}\left(\widetilde{S}(\boldsymbol{\theta},\boldsymbol{\delta})\right)$ is CVaR at the confidence level of α. The objective function controls the trade-off between the expectation and CVaR with a parameter $\gamma\in[0, 1]$. Setting $\gamma = 0$ yields the totally risk-averse case. Increasing the value of γ makes the strategy more risk-neutral, with eventually $\gamma = 1$ representing the completely risk-neutral case. More specifically, the revenue $\widetilde{S}(\boldsymbol{\theta},\boldsymbol{\delta})$ can be expressed as

$$\widetilde{S}(\boldsymbol{\theta},\boldsymbol{\delta}) = \sum_{t=1}^{T}\left[\widetilde{\pi}_t^{da}p_t^{bid} + \widetilde{\pi}_t^{rt}\left(p_t^{out} - p_t^{bid}\right)\right] + \widetilde{\pi}^E\Delta E \tag{4.6}$$

where p_t^{out} is the total output of the REG–ESS union at time interval t and $p_t^{out} = [1, -1, 1]\boldsymbol{u}_t$. The revenue consists of three parts. The first part $\widetilde{\pi}_t^{da}p_t^{bid}$ is the revenue from day-ahead offering, while the second part $\widetilde{\pi}_t^{rt}\left(p_t^{out} - p_t^{bid}\right)$ is the revenue or penalty from the balancing market. The last part is the so-called *energy value*. Unlike some published work where residual energy of ESS at the end of the scheduling horizon is either overlooked or anchored within an acceptable deviation from the initial level [12], the concept of *energy value* is used here to reflect the value of the residual energy in ESS, since the residual energy has the potential to get revenues by generating (discharging) in following horizons. It is inspired by an essential concept, *water value* in hydro-power management, which relates to the marginal cost of water in the reservoirs, and is equal to the replacement value of other general generators or average market prices in ideal cases. *Energy value* introduced here is used to reflect the influence of residual energy on the daily revenue. Consequently, for the sake of simplicity, the final residual energy deviation of ESS, $\Delta E = x_T - x_0$, is assigned a price, $\widetilde{\pi}^E$, to reflect its value, and the resulting energy value is considered in the objective function (4.6). Substituting Eq. (4.2) into (4.6), the stochastic revenue function can be reformulated as

$$\widetilde{S}(\boldsymbol{\theta},\boldsymbol{\delta}) = \widetilde{\boldsymbol{\pi}}_t^{rt}\left[\left(\boldsymbol{D}^{rg} + \boldsymbol{D}^{sd} - \boldsymbol{D}^{sc}\right)\boldsymbol{\delta} + \left(\hat{\boldsymbol{P}}^{rg} + \hat{\boldsymbol{P}}^{sd} - \hat{\boldsymbol{P}}^{sc}\right)\right] + \left(\widetilde{\boldsymbol{\pi}}^{da} - \widetilde{\boldsymbol{\pi}}^{rt}\right)\boldsymbol{p}^{bid} + \widetilde{\pi}^E\Delta E \tag{4.7}$$

where $\widetilde{\boldsymbol{\pi}}^{da}$ and $\widetilde{\boldsymbol{\pi}}^{rt}$ are vectors of $\widetilde{\pi}_t^{da}$ and $\widetilde{\pi}_t^{rt}$.

CVaR is defined in reference [13] as follows:

$$\mathrm{CVaR}_{\alpha} = \mathbb{E}\{s \mid s \le \mathrm{VaR}_{\alpha}(s)\} = \frac{1}{\alpha}\int_0^{\mathrm{VaR}_{\alpha}(s)} sp(s)ds \tag{4.8}$$

where $\widetilde{S}(\boldsymbol{\theta}, \boldsymbol{\delta})$ is denoted by s for convenience, $p(s)$ is the probability density function of the random variable s, and $\text{VaR}_\alpha(s)$ is defined as

$$\text{VaR}_\alpha(s) = \min\{V \in \mathbb{R} : P(s \leq V) \geq \alpha\} \tag{4.9}$$

where $P(s)$ is the cumulative distribution function of s.

The above definition of VaR makes it difficult to calculate CVaR. Inspired by the work of reference [14], a simpler function is defined as follows:

$$F(s, v, \alpha) = v + \frac{1}{\alpha} \int_{s < v} (s - v)p(s)ds \tag{4.10}$$

whose maximum w.r.t. v can be used as the CVaR. A detailed and strict proof can be found in references [15] and [16]. There follows a simple explanation. The derivative of Eq. (4.10) w.r.t. v is

$$\frac{\partial F(s, v, \alpha)}{\partial v} = 1 - \frac{1}{\alpha} \int_{s < v} p(s)ds$$

By equating it to 0, it can be found that VaR maximizes $F(s, v, \alpha)$ w.r.t. v (the second-order derivative of F w.r.t. v as $(1/\alpha)p(v)$ is negative). Then, according to the definition of CVaR, one can see that the maximum of function $F(s, v, \alpha)$ w.r.t. v is equal to CVaR.

For the optimization problem of this section, uncertain prices and wind power output in the objective function are represented by scenarios. Define Ω as the set of scenarios, ρ_ω as the probability for scenario ω, and $\sum_{\omega \in \Omega} \rho_\omega = 1$; then the expectation can be formulated as

$$\mathbb{E}\left(\widetilde{S}(\boldsymbol{\theta}, \boldsymbol{\delta})\right) = \sum_{\omega \in \Omega} \rho_\omega S_\omega \tag{4.11}$$

where

$$S_\omega = \pi_\omega^{rt}\left[\left(\mathbf{D}^{rg} + \mathbf{D}^{sd} - \mathbf{D}^{sc}\right)\boldsymbol{\delta}_\omega + \left(\hat{\boldsymbol{P}}^{rg} + \hat{\boldsymbol{P}}^{sd} - \hat{\boldsymbol{P}}^{sc}\right)\right] + \left(\pi_\omega^{da} - \pi_\omega^{rt}\right)\boldsymbol{p}^{bid} + \pi_\omega^E \Delta E_\omega$$

and π_ω^{rt} is the realization of the real-time price under scenario ω, while the same goes for the variables such as $\boldsymbol{\delta}_\omega$, π_ω^{da}, π_ω^E, and ΔE_ω. Equation (4.10) can be discretized as [11]

$$\text{CVaR}_\alpha\left(\widetilde{S}(\boldsymbol{\theta}, \boldsymbol{\delta})\right) = v + \frac{1}{\alpha}\sum_{\omega \in \Omega} \rho_\omega[S_\omega - v]^- \tag{4.12}$$

where $[x]^- = \min\{x, 0\}$. By introducing the slack variables $\{z_\omega\}$, Eq. (4.12) can be linearized as

$$
\begin{cases}
\mathrm{CVaR}_\alpha\left(\widetilde{S}(\theta,\delta)\right) = v + \dfrac{1}{\alpha}\displaystyle\sum_{\omega\in\Omega}\rho_\omega z_\omega \\[2mm]
z_\omega \leq S_\omega - v \\[1mm]
z_\omega \leq 0
\end{cases}
\tag{4.13}
$$

Substituting Eqs. (4.11) and (4.13) into Eq. (4.4), the objective function can be reformulated linearly as

$$
\max_{\theta}\ \gamma\sum_{\omega\in\Omega}\rho_\omega S_\omega + (1-\gamma)\left(v + \frac{1}{\alpha}\sum_{\omega\in\Omega}\rho_\omega z_\omega\right)
\tag{4.14}
$$

$$
\text{s.t.}\ \ S_\omega = \pi_\omega^{rt}\left[\left(\boldsymbol{D}^{rg} + \boldsymbol{D}^{sd} - \boldsymbol{D}^{sc}\right)\delta_\omega + \left(\hat{\boldsymbol{P}}^{rg} + \hat{\boldsymbol{P}}^{sd} - \hat{\boldsymbol{P}}^{sc}\right)\right] + \left(\pi_\omega^{da} - \pi_\omega^{rt}\right)\boldsymbol{p}^{bid}
$$

$$
+ \pi_\omega^E \Delta E_\omega, \forall \omega \in \Omega
\tag{4.15}
$$

$$
z_\omega \leq S_\omega - v, z_\omega \leq 0
\tag{4.16}
$$

4.3.2 Constraints

The constraint set (4.5) is built to guarantee the feasibility of solutions for any realization of δ within a certain neighborhood of the forecast value $\boldsymbol{d}_0$. The uncertainty set is defined as $\boldsymbol{H}\delta \leq \boldsymbol{h}$, where $\boldsymbol{H} = [\boldsymbol{I}_{3T}, -\boldsymbol{I}_{3T}]^{\mathrm{T}}$, $\boldsymbol{h} = \left[\delta^+\, \boldsymbol{1}_{3T}^{\mathrm{T}}, -\delta^-\, \boldsymbol{1}_{3T}^{\mathrm{T}}\right]^{\mathrm{T}}\boldsymbol{d}_0$, δ^+ and δ^- are up and down deviation factors of the forecast values, and $\boldsymbol{I}_{3T}$, $\boldsymbol{1}_{3T}$ are a 3T-dimensional unit matrix and a $3T \times 1$ column vector with all elements set to 1, respectively. Constraints in Eq. (4.5) include

$$
\boldsymbol{1}_T E_{\min} \leq \boldsymbol{x}(\theta,\delta) \leq \boldsymbol{1}_T E_{\max}
\tag{4.17}
$$

$$
\boldsymbol{1}_T \underline{p}^{sc} \leq \boldsymbol{D}^{sc}\delta + \hat{\boldsymbol{p}}^{sc} \leq \boldsymbol{1}_T \overline{p}^{sc}
\tag{4.18}
$$

$$
\boldsymbol{1}_T \underline{p}^{sd} \leq \boldsymbol{D}^{sd}\delta + \hat{\boldsymbol{p}}^{sd} \leq \boldsymbol{1}_T \overline{p}^{sd}
\tag{4.19}
$$

$$
0 \leq \boldsymbol{D}^{rg}\delta + \hat{\boldsymbol{p}}^{rg} \leq \boldsymbol{p}^{rf}
\tag{4.20}
$$

$$
0 \leq \boldsymbol{p}^{bid} \leq \boldsymbol{1}_T C^{rg}
\tag{4.21}
$$

Equation (4.17) limits the residual energy of ESS at each interval within allowable ranges $[E_{\min}, E_{\max}]$. The same goes for the charging and discharging power of ESS, as in Eqs. (4.18) and (4.19), with lower and upper levels of charging and discharging power $\left[\underline{p}^{sc}, \overline{p}^{sc}\right]$ and $\left[\underline{p}^{sd}, \overline{p}^{sd}\right]$, respectively. Moreover, considering the possibility of curtailing REG output, Eq. (4.20) indicates that the actual wind power output will not exceed the real-time forecast value, $\boldsymbol{p}^{rf}$. The offered amount should be nonnegative as the role of REG–ESS is as a generator in the market, and the amount should be below the maximum allowable injection power of the

REG–ESS union, C^{rg}, given by Eq. (4.21). In references such as [17], components of the decision matrix D, such as D_{rt}^{sc}, are set as lower triangular matrices as future information is not available at any given time. Here, future information is substituted with updated forecast values for the balancing price and REG power-related matrices. Hence, the corresponding matrices do not have to be lower-triangulars. Moreover, the day-ahead price-related matrices, such as D_{da}^{sc}, are not lower triangulars as day-ahead prices have already been determined before operation.

Constraint (4.17) can be formulated as

$$
\begin{bmatrix} I_T \\ -I_T \end{bmatrix} \begin{pmatrix} x_0 \mathbf{1}_T + I_{dr,T}\left(\eta_c \hat{\boldsymbol{p}}^{sc} - \dfrac{1}{\eta_d}\hat{\boldsymbol{p}}^{sd}\right) \\ + I_{dr,T}\left(\eta_c \boldsymbol{D}^{sc} - \dfrac{1}{\eta_d}\boldsymbol{D}^{sd}\right)\delta \end{pmatrix} \leq \begin{bmatrix} \mathbf{1}_T E_{max} \\ -\mathbf{1}_T E_{min} \end{bmatrix} \tag{4.22}
$$

where $I_{dr,\,T}$ is a T-dimensional lower triangular matrix with all elements as 1. Similarly, constraints (4.18) to (4.20) can be reformulated as

$$
\begin{bmatrix} I_T \\ -I_T \end{bmatrix}\hat{\boldsymbol{p}}^{sc} + \begin{bmatrix} I_T \\ -I_T \end{bmatrix}\boldsymbol{D}^{sc}\delta \leq \begin{bmatrix} \mathbf{1}_T \overline{p}^{sc} \\ -\mathbf{1}_T \underline{p}^{sc} \end{bmatrix} \tag{4.23}
$$

$$
\begin{bmatrix} I_T \\ -I_T \end{bmatrix}\hat{\boldsymbol{p}}^{sd} + \begin{bmatrix} I_T \\ -I_T \end{bmatrix}\boldsymbol{D}^{sd}\delta \leq \begin{bmatrix} \mathbf{1}_T \overline{p}^{sd} \\ -\mathbf{1}_T \underline{p}^{sd} \end{bmatrix} \tag{4.24}
$$

$$
\begin{bmatrix} I_T \\ -I_T \end{bmatrix}\hat{\boldsymbol{p}}^{rg} + \begin{bmatrix} \boldsymbol{D}^{rg} - [0,0,I_T] \\ -\boldsymbol{D}^{rg} \end{bmatrix}\delta \leq \begin{bmatrix} [0,0,I_T]d_0 \\ 0 \end{bmatrix} \tag{4.25}
$$

Furthermore, constraints (4.22) to (4.25) can be reformulated in order to eliminate the random variable δ and to have finite cardinality through duality theory, as performed by reference [17]. More specifically, these constraints can be presented in a concise way as $\boldsymbol{c}^{\mathsf{T}}\delta \leq q$ with δ satisfying $H\delta \leq h$. This yields

$$
\max_{\delta} \ \left\{ \begin{array}{l} \boldsymbol{c}^{\mathsf{T}}\delta, \\ s.t. \ \ H\delta \leq h : \boldsymbol{\mu} \end{array} \right\} \leq q \tag{4.26}
$$

where $\boldsymbol{\mu}$ is the dual variable associated with the constraints. According to duality theory, Eq. (4.26) is equivalent to

$$
\left\{ \begin{array}{l} \boldsymbol{\mu}^{\mathsf{T}}h \leq q \\ \boldsymbol{\mu}^{\mathsf{T}}H = \boldsymbol{c} \\ \boldsymbol{\mu} \geq 0 \end{array} \right. \tag{4.27}
$$

4.3.3 Complete Optimization Formulation

Now, the whole optimization problem can be formulated in a linear way as follows:

$$\max (4.14)$$
$$\text{s.t. } (4.15), (4.16), (4.21)$$
$$(4.22)–(4.25) \text{ in the form of } (4.27)$$

It should be noted that the above formulation is based on the one-price balancing market (see the discussions in Chapter 3), while this can be readily extended to the two-price case. This is done by replacing the revenue in Eq. (4.6) with the following one:

$$
\begin{cases}
\widetilde{S}(\boldsymbol{\theta}, \boldsymbol{\delta}) = \sum\limits_{t=1}^{T} \left[\widetilde{\pi}_t^{da} p_t^{bid} - \widetilde{\pi}_t^{rt\,+} p_t^{up} + \widetilde{\pi}_t^{rt\,-} p_t^{dw} \right] + \widetilde{\pi}^E \Delta E \\
p_t^{dw} - p_t^{up} = p_t^{out} - p_t^{bid} \\
p_t^{dw}, p_t^{up} \geq 0
\end{cases}
\tag{4.28}
$$

where $\widetilde{\pi}_t^{rt\,+}$ and $\widetilde{\pi}_t^{rt\,-}$ are up-regulation and down-regulation prices and p_t^{dw}, p_t^{up} are positive and negative deviations between actual output and day-ahead offers, respectively. Either p_t^{dw} or p_t^{up} should be zero due to the optimization requirement. This characteristic ensures that p_t^{dw} and p_t^{up} can precisely represent the required down-regulation and up-regulation power capacity, respectively.

4.3.4 Case Studies

Case studies are based on realistic data from the Nord Pool market [18] and wind farms in Denmark [19, 20]. Per-unit data of wind power forecast scenarios are provided by [21] and translated into actual data by multiplying C^{rg}. One hundred scenarios are applied to optimization, and are also used to fit distributions of wind power at each time interval. Then quantities of wind power scenarios for strategy evaluation and sensitivity analyses are generated based on these distributions, as performed in reference [12].

Day-ahead prices and up/down-regulation prices are from the DK-West area in the Nord Pool market during January 1 to 10, 2014. Because the Nord Pool is of a two-price balancing market, up/down-regulation prices are different and one or the other of them is equal to the day-ahead price at any specific time interval. Therefore, we take the different one as the balancing price in the one-price balancing market. Similarly, price scenarios are necessary for both optimization and strategy evaluation, and are generated by

Table 4.1 Parameters of energy storage system and wind power capacity.

η_c	η_d	E_{min}(WMh)	E_{max}(WMh)
0.95	0.95	10	50

E_0(WMh)	$\overline{p}^{sc}$(WM)	$\overline{p}^{sd}$(WM)	C^{rg}(WM)
30	10	10	100

$$\begin{cases} \widetilde{\pi}_t^{da} = \overline{\pi}_t^{da}\left(1 + \sigma^{da}\widetilde{\varepsilon}\right) \\ \widetilde{\pi}_t^{rt} = \overline{\pi}_t^{rt}(1 + \sigma^{rt}\widetilde{\varepsilon}) \end{cases}$$

where $\overline{\pi}_t^{da}, \overline{\pi}_t^{rt}$ are actual data of the day-ahead and balancing prices at time interval t, $\widetilde{\varepsilon}$ is a random variable that obeys the standard normal distribution $\widetilde{\varepsilon} \sim N(0, 1)$, and σ^{da} and σ^{rt} are the standard deviations for day-ahead and balancing prices, respectively. The *energy value* of the ESS is updated every day and set as the average spot price, as stated in papers in the field of hydropower [22]. Parameters of the ESS are listed in Table 4.1. In the following case studies, key parameters are set as $\gamma = 0.9$, $\Delta^+ = \Delta^- = 0.1$, $\sigma^{da} = 0.2$, $\sigma^{rt} = 0.3$, and $\alpha = 0.05$.

4.3.4.1 An Illustrative Case
In this case, only the first two time intervals of wind power and price data are considered. The optimal solution includes

$$\hat{p}^{rg} = [52.9, 59.3]^T, \hat{p}^{sc} = [5, 5]^T, \hat{p}^{sd} = [5, 5]^T,$$

$$D_{rt}^{sc} = \begin{bmatrix} -2.17 & 0 \\ 0 & -2.34 \end{bmatrix}, D_{rt}^{sd} = \begin{bmatrix} 0.15 & -2.17 \\ -2.17 & 0 \end{bmatrix}, D_{rf}^{rg} = \begin{bmatrix} 1 & 0 \\ 0 & 1 \end{bmatrix}$$

while $D_{da}^{sc}, D_{wf}^{sc}, D_{da}^{sd}, D_{rf}^{sd}, D_{da}^{rg}$, and D_{rt}^{rg} are all zeros. From the results it can be seen that scheduled wind power is independent of prices, while charging and discharging power of ESS only depend on balancing prices. Assume that the wind power forecast error $\Delta p^{rg} = [6, 7]$ MW and day-ahead price forecast error $\Delta \pi^{da} = [1, 2]$ EUR MWh^{-1}, while the balancing price forecast error $\Delta \pi^{rt} = [-1, 1]$ EUR MWh^{-1}. According to Eq. (4.3), decisions of wind power and charging/discharging power for each interval can be calculated as

$$p_1^{rg} = 52.9 + 1 \times 6 = 58.9 \text{ MW}$$

$$p_2^{rg} = 59.3 + 1 \times 7 = 66.3 \text{ MW}$$

$$p_1^{sc} = 5 + (-2.17) \times (-1) = 7.17 \text{ MW}$$

$$p_2^{sc} = 5 + (-2.34) \times 1 = 2.66 \text{ MW}$$

$$p_1^{sd} = 5 + 0.15 \times (-1) + (-2.17) \times 1 = 2.68 \text{ MW}$$

$$p_2^{sd} = 5 + (-2.17) \times (-1) = 7.17 \text{ MW}$$

Since simultaneous charging and discharging is prohibited and $p_1^{sc} > p_1^{sd}$, the actual discharging power of the first time interval is zero, while the actual charging power is $p_1^{sc} - p_1^{sd} = 4.49$ MW. Then in the second time interval, as $p_2^{sc} < p_2^{sd}$, the actual charging power of this interval is zero, while the actual discharging power is $p_1^{sd} - p_1^{sc} = 4.51$ MW. The power decision procedure can be extended to cases with more time intervals.

4.3.4.2 Comparisons of Revenue by Different Strategies

The revenue earned by the proposed strategy is compared to some other adopted strategies. Strategy 1 is used as a benchmark, which bids the forecast value and generates without any curtailments for a wind farm operating alone. Strategy 2 is the *Expected Utility Maximization* (EUM) strategy, where the wind farm bids the optimal quantile of wind power forecast and generates without any curtailment [23]. In both strategies, the wind farm works alone without the ESS. Strategy 3 is the *Filter Control Strategy* (FCS) [24], where the ESS is utilized to compensate for the deviations between wind power output and day-ahead bidding to reduce the deviation penalty cost. Strategy 4 is the strategy derived above based on linear decision rules.

The comparisons are carried out using 10 000 test scenarios in a Monte Carlo simulation framework. The distributions of revenue over all the scenarios for the strategies are demonstrated in Figure 4.3, while the mean values of revenues obtained by the four strategies are also listed within the legend of the figure. Several conclusions can be reached from the comparison. Firstly, when the WF operates alone, the EUM strategy can help to increase the revenue (compare Strategy 1 and Strategy 2). Secondly, the FCS does not have obvious advantages over the EUM strategy, which confirms the conclusion in reference [12] that only with proper control strategy can ESS bring satisfactory benefit. Thirdly, the control strategy based on linear decision rules has the best performance as the probability distribution of its revenue shifts right, which obviously increases the revenue of the wind farm by over 11%.

4.4 Electricity Market Time Frame and Rules with Intraday Market

Except for the day-ahead and real-time markets, intraday markets are considered in this section. In order to better accommodate the day-ahead and intraday

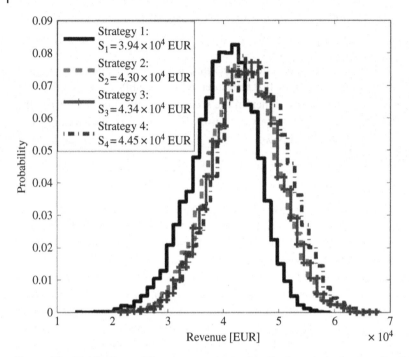

Figure 4.3 Distributions of the revenues obtained by the four strategies based on 10 000 scenarios.

combined strategy, an electricity market rule including three stages, i.e. day-ahead, intraday, and real-time stages, will be discussed. The optimal bidding and operation strategies of the REG–ESS union will be based on this three-stage market rule in the following section of this chapter. As depicted in Figure 4.4, the day-ahead market covers the 24 hours of day $(D + 1)$ [25]. Each intraday market only covers four hours. It should be noted that market participants may consider the time span covering the following 24 hours in each intraday energy rebidding, but only submit the biddings of the next four hours to the system operator. In real-time operation, market participants try to follow their biddings and get a settlement according to their actual output.

4.4.1 Day-Ahead Bidding Rules

It is assumed that the day-ahead market begins at 12:00 every day. Power producers submit their biddings to the market with 12–36 hours in advance [9]. Take an REG–ESS union, for example, which needs to submit the power bids $\{p_t^{da}\}$ for

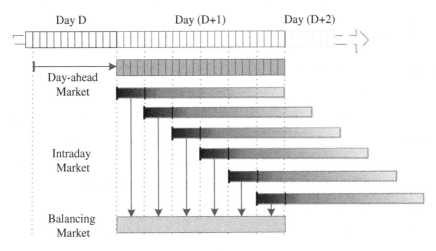

Figure 4.4 Time frame illustration of the three electricity markets at different stages.

each of the 24 hours of the next market day, as illustrated in reference [26]. The price bids will not be considered as we assume the REG–ESS union is a price-taker.

4.4.2 Intraday Bidding Rules

In order to better utilize the updated REG power forecast, the following intraday bidding rules are assumed:

1) In the intraday market, participants can submit their biddings every four hours. Namely, the intraday market opens at 0:00, 4:00, 8:00, 12:00, 16:00, and 20:00 within each day.
2) At the intraday bidding stage, power producers can revise their generation plan for the future 4–8 hours, at the time resolution of 10 minutes. The deviation between intraday and day-ahead biddings will be penalized.
3) The submitted generation plan cannot be revised any more and participants would be penalized if a deviation occurred in the real-time operation.

4.4.3 Real-Time Operation

For the real-time operation, power producers will be penalized if there is a deviation between the intraday bidding and the average power output of every 10 minutes. Besides, the fluctuation of output should be constrained according to corresponding technical standards. For example, in China, the allowable wind power change in 1 minute and 10 minutes is determined by the installed capacity

Table 4.2 Standard for wind power variation limit.

Capacity (MW)	Limit of active power variation in 10 min (MW)	Limit of active power variation in 1 min (MW)
<30	10	3
30–150	Capacity/3	Capacity/10
>150	50	15

of the wind farm, as shown in Table 4.2. In Germany, the fluctuation in 1 minute should be below 10% of the wind farm capacity [27].

Supervisory control and data acquisition (SCADA) system of a wind farm can record data at a resolution higher than 1 minute. So the active power change in 1 minute, $\Delta p_{k,1}$, can be obtained easily, as shown in Eq. (4.29). However, the definition of power change in 10 minutes is ambiguous. Two definitions, $\Delta p_{k,10}^{I}$ and $\Delta p_{k,10}^{II}$, given by Eqs. (4.30) and (4.31), respectively, are considered:

$$\Delta p_{k,1} = \left| p_{k+1} - p_k \right| \tag{4.29}$$

$$\Delta p_{k,10}^{I} = \left| P_{k+10} - P_k \right| \tag{4.30}$$

$$\Delta p_{k,10}^{II} = \max \left\{ \left(P_i - P_j \right) \mid i,j = k, k+1...k+9 \right\} \tag{4.31}$$

To sum up, at 12:00 of Day D, the day-ahead market opens and the REG–ESS union can bid its generation plan for 0:00–24:00 of Day $(D+1)$. It can also bid a generation plan for a future four to eight hours, i.e. 16:00–20:00. At 16:00, the intraday market opens again and the union can bid its generation plan for 20:00–24:00. The same process occurs every four hours. With time approaching, the revenues and penalties are settled every 10 minutes according to the average power within the corresponding 10 minutes and the intraday biddings.

4.5 Rolling Optimization Framework and Mathematical Formulations Considering Intraday Markets

4.5.1 Data Flow among Different Sections

Operation decisions of the REG–ESS union in one section of a market couple with those of other sections, and together they affect the total revenue. Figure 4.5 illustrates the data flow between different optimization modules. Taking an REG power forecast for the future 12–36 hours and an initial residue energy forecast of ESS as input, the REG–ESS union can submit the day-ahead biddings based

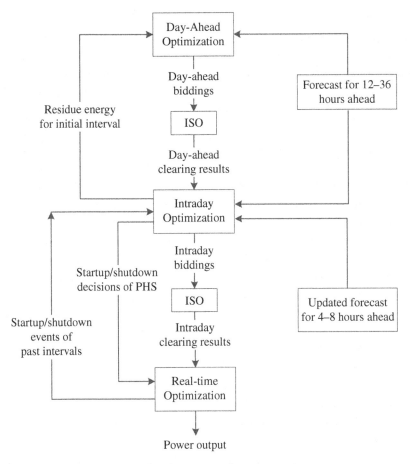

Figure 4.5 Diagram of data flow between each market section.

on the day-ahead optimization model. After the day-ahead market clearing is completed, the results, including cleared electricity quantities and prices, are passed to the intraday optimization module, whose output is used to calculate the expected revenue. In the following discussions, it is assumed that the REG–ESS union is a price taker and all its biddings are accepted by the electricity market. The real-time optimization module makes the final output decisions based on the cleared intraday biddings and charging/discharging decisions, and the actual revenue can be computed. If the ESS is a pumped hydroelectric storage (PHS) plant, the completed startup and shutdown events are counted and considered in the intraday optimizations (as illustrated in Figure 4.5).

4.5.2 Initial Residue Energy of Different Optimizations

The most notable difference between an ESS and a thermal generator is that the limited energy capacity of ESS makes its output/input power of different time intervals coupled with each other. For this reason, the initial energy of ESS is essential in the optimization model. This is because if the actual initial energy of ESS deviated much from what was set in the optimization model, the residue energy in following time intervals may violate the allowable ranges according to the operational schedule. The initial energy of ESS used for the rolling optimization is determined and adjusted as follows:

1) In day-ahead optimization, the initial SoC value of day D, $E_{0,D}^{da}$, is determined by the intraday optimization conducted at 12:00 and is equal to the optimal value at 24:00 day $(D\text{-}1)$, $E_{24,D-1}^{id,12}$.

2) The kth intraday optimization begins at the $(4k-4)$th hour. The initial residue energy value, $E_{4k,D}^{id,4k-4}$, is determined by the result of the $(k-1)$th optimization. More specifically, the value $E_{4k,D}^{id,4k-4}$ is equal to the sum of the $(k-1)$th intraday optimization result, $E_{4k,D}^{id,4k-8}$, and the latest adjustment δ where δ is the deviation of real residue energy $E_{4k-4,\,D}$ and initial residue energy $E_{4k-4,D}^{id,4k-8}$ of the $(k-1)$th intraday optimization $\delta = E_{4k-4,D} - E_{4k-4,D}^{id,4k-8}$, $E_{4k,D}^{id,4k-4} = E_{4k-4,D} - E_{4k-4,D}^{id,4k-8} + E_{4k,D}^{id,4k-8}$.

3) In real-time optimal control, the initial residue energy of the current control period is just the present measurement value.

4.5.3 Optimization Model for Each Market

4.5.3.1 Nomenclature

In this section, the REG-ESS union composed of wind farm(s) and a PHS plant will be considered. It should be noted that the models can be tailored without difficulty to deal with the REG-ESS union composed of solar power plant(s) and/or other types of ESS. To aid understanding, nomenclature is given in the following table:

Index sets:

$I = \{1, 2, ..., s\}$	Set of day-ahead wind power profile scenarios
$H = \{1, 2, ..., 24\}$	Set of hourly periods in the future 12–36 hours
$K = \{1, 2, ..., 24\}$,	Set of 10-min intervals in the future 4–8 hours
$J = \{1, 2, ..., 20\}$	Set of hourly periods in the future 8–28 hours
$M = \{1, 2, ..., 10\}$	Set of minutely intervals in the future 1–10 minutes

Functions:

$Pr(\cdot)$	Probability of an event
$Ex(\cdot)$	Expectation of an stochastic variable
$S_t^K(\cdot)$	Revenue in interval t of set K

$S_t^J(\cdot)$	Revenue in interval t of set J
$V(\cdot)$	Maximum variation of a sequential data

Variables:

p_t^{da}/p_t^{ha}	Day-ahead/intraday power bidding for interval t
$p_{t,i}^w$	Power generation of wind farm (WF) in interval t, scenario i
$p_{t,i}^p/p_{t,i}^g$	Pumping/generating power of pumped hydroelectric storage plant at interval t, scenario i
$s_{t,i}^p/s_{t,i}^g$	On–off state of pumping/generating at interval t, scenarios i
n_t^p	Number of running pumping turbines at interval t
n_t^{su}/n_t^{sd}	Number of startup/shutdown pumping turbines at the end of interval t
$E_{t,i}$	Residue energy of PHS to generate at interval t, scenario i
ω_t^p	Weighting factor of revenue in objective function of interval t
ω^E	Weighting factor of residue energy deviation in objective function of interval t
$\Delta p_t^+/\Delta p_t^-$	Positive/negative deviation between actual output and intraday bidding of interval t
Δp_t^{bid}	Deviation between day-ahead and intraday bidding of interval t
$p_{t,a}^{union}/p_{t,ins}^{union}$	Actual power/output instruction of WF–PHS union at interval t

Constants:

ρ_i	Weighting factor of wind power scenario i in the day-ahead optimization
Δt_m	Duration of an interval in index set K
Δt_h	Duration of an interval in index set J, H
π_t	Electricity price of interval t
E_{min}/E_{max}	Lower/upper bound of residue energy of PHS
α	Confidence level of chance constraint
$p_{t,i}^{wf}$	Forecasted wind power of interval t, scenario i
C_{su}/C_{sd}	Startup/shutdown cost of pumping unit per event
γ^+/γ^-	Penalty factor of electricity price for positive/negative deviation between actual output and bidding of interval t
β	Penalty factor of electricity price for the deviation between day-ahead and intraday bidding of interval t
B_f^1/B_f^{10}	Variation limit on wind power output for every 1 minute/10 minutes
N	Total number of pumping or generating turbines
N_{ud}	Upper limit of the total startup and shutdown times per day for pumping turbines
p_{min}/p_{max}	Minimal/maximal pumping power of each unit
g_{min}/g_{max}	Minimal/maximal generating power of each unit
η_c/η_d	Pumping/generating efficiency of the PHS

4.5.3.2 Day-Ahead Market

In the day-ahead market, variables including $\left\{ p_t^{da}, p_{h,i}^w, p_{h,i}^g, p_{h,i}^p, n_{h,i}^{su}, n_{h,i}^{sd} \right\}$ should be determined through optimization but only the day-ahead biddings $\left\{ p_t^{da} \right\}$ will be passed to the intraday optimization model. The day-ahead optimization model is a two-stage model. The day-ahead biddings are *here-and-now solutions*, which make the expected revenue maximal, and the other intermediate ones are *wait-and-see* solutions. The uncertainty of wind power is represented by multiple scenarios, which are obtained based on the forecast results:

$$
\max \sum_{i \in I} \left[\rho_i \sum_{h \in H} \left(\begin{array}{c} \pi_h \left(p_{h,i}^w + p_{h,i}^g - p_{h,i}^p \right) \Delta t_h - \left(C_{su} n_{h,i}^{su} + C_{sd} n_{h,i}^{sd} \right) \\ -\gamma^+ \pi_h \Delta p_{h,i}^+ \Delta t_h - \gamma^- \pi_h \Delta p_{h,i}^- \Delta t_h \end{array} \right) \right] \tag{4.32}
$$

$$
\text{s.t.} \ 0 \le p_{h,i}^w \le p_{h,i}^{wf}, \quad h \in H, i \in I \tag{4.33}
$$

$$
\Delta p_{h,i}^+ = \max \left(0, p_{h,i}^w + p_{h,i}^g - p_{h,i}^p - p_h^{da} \right), \quad h \in H, i \in I \tag{4.34}
$$

$$
\Delta p_{h,i}^- = -\min \left(0, p_{h,i}^w + p_{h,i}^g - p_{h,i}^p - p_h^{da} \right), \quad h \in H, i \in I \tag{4.35}
$$

$$
s_{h,i}^p + s_{h,i}^g \le 1, \quad s_{h,i}^p, s_{h,i}^g \in \{0,1\}, h \in H, i \in I \tag{4.36}
$$

$$
s_{h,i}^p \le n_{h,i}^p \le s_{h,i}^p N, \quad h \in H, i \in I \tag{4.37}
$$

$$
n_{h+1,i}^p = n_{h,i}^p + n_{h,i}^{su} - n_{h,i}^{sd}, \quad h \in H, i \in I \tag{4.38}
$$

$$
p_{\min} n_{h,i}^p \le p_{h,i}^p \le p_{\max} n_{h,i}^p, \quad h \in H, i \in I \tag{4.39}
$$

$$
s_{h,i}^g g_{\min} \le p_{h,i}^g \le s_{h,i}^g g_{\max} N, \quad h \in H, i \in I \tag{4.40}
$$

$$
\sum_{h=1}^{24} \left(n_{h,i}^{su} + n_{h,i}^{sd} \right) \le N_{ud}, \quad h \in H, i \in I \tag{4.41}
$$

$$
E_{h+1,i} = E_{h,i} + p_{h,i}^p \Delta t_h \eta_c - p_{h,i}^g \Delta t_h / \eta_d, \quad h \in H, i \in I \tag{4.42}
$$

$$
E_{\min} \le E_{h,i} \le E_{\max}, \quad h \in H, i \in I \tag{4.43}
$$

Equation (4.32) considers the revenue of generation, costs of turbine startup and shutdown, and the penalty for power deviation. Constraints (4.33) limit the output range of the wind farm. Positive and negative deviations are defined in Eqs. (4.34) and (4.35). Constraints (4.36) prohibit the simultaneous charging (pumping) and discharging (generating) of the reversible turbines. On/off states and number of pumping turbines are restricted by constraints (4.37) to (4.41). Constraints (4.42) and (4.43) count the residue energy and confine it within an allowable range of the PHS.

4.5.3.3 Intraday Market

Although an REG–ESS union only needs to bid for the future 4–8 hours every time the intraday market opens, the wind power of the future hours should be considered because the residue energy of ESS is coupled with the *farther future*. On the other side, as the wind power forecast error increases distinctly with time, the *near* and *farther future* decision should be set with different weights. Consequently, it is better to define the objective function and constraints of these two parts separately. For the *near future*, as the forecast errors are relatively small, a chance constraint is applied. For the *farther future*, scenarios are generated to represent the stochastic nature of wind power as the chance constraints for the large forecast error case would narrow the feasible region, which would reduce the expected revenue.

The detailed formulation for the intraday market decision is described as follows:

$$\max S \tag{4.44}$$

$$\text{s.t.} \quad S = \sum_{t \in K} S_t^K + \sum_{t \in J} \omega_t^p S_t^J \tag{4.45}$$

$$S_t^K = \pi_t \left(p_t^w + p_t^g - p_t^p \right) \Delta t_m - \left(C_{su} n_t^{su} + C_{sd} n_t^{sd} \right) - \beta \pi_t \Delta p_t^{bid} \Delta t_m, \quad t \in K \tag{4.46}$$

$$S_t^J = \text{Ex} \left\{ \begin{array}{l} \pi_t \left(p_{t,i}^w + p_{t,i}^g - p_{t,i}^p \right) \Delta t_h \\ - \pi_t \left(\gamma^+ \Delta p_{t,i}^+ + \gamma^- \Delta p_{t,i}^- \right) \Delta t_h \end{array} \right\} - \left(C_{su} n_t^{su} + C_{sd} n_t^{sd} \right) - \beta \pi_t \Delta p_t^{bid} \Delta t_h,$$

$$t \in J, i \in I \tag{4.47}$$

$$p_t^w + p_t^g - p_t^p = P_t^{ha}, \quad t \in K \tag{4.48}$$

$$\Pr \left\{ 0 \le p_t^w \le p_t^{wf} + \xi_t(v_t) \right\} > \alpha, \quad t \in K \tag{4.49}$$

$$-\Delta p_t^{bid} \le P_t^{ha} - p_t^{da} \le \Delta p_t^{bid}, \quad t \in K, J \tag{4.50}$$

$$-\Delta p_{t,i}^- \le p_{t,i}^w + p_{t,i}^g - p_{t,i}^p - p_t^{ha} \le \Delta p_t^+, \quad t \in J, i \in I \tag{4.51}$$

$$\left| p_t^{ha} - p_{t-1}^{ha} \right| \le B_f^{10}, \quad t \in K \tag{4.52}$$

$$s_t^p + s_t^g \le 1, \quad s_t^p, s_t^g \in \{0,1\}, \quad t \in K, J \tag{4.53}$$

$$s_t^p \le n_t^p \le s_t^p N, \quad t \in K, J \tag{4.54}$$

$$n_{t+1}^p = n_t^p + n_t^{su} - n_t^{sd} \quad t \in K, J \tag{4.55}$$

$$\sum_{t \in K, J} \left(n_t^{su} + n_t^{sd} \right) < N_{ud} \tag{4.56}$$

$$p_{\min} n_t^p \le p_t^p \le p_{\max} n_t^p, \quad t \in K \tag{4.57}$$

$$p_{\min} n_{t,i}^p \le p_{t,i}^p \le p_{\max} n_{t,i}^p, \quad t \in J, i \in I \tag{4.58}$$

$$s_t^g g_{\min} \le p_t^g \le s_t^g g_{\max} N, \quad t \in K \tag{4.59}$$

$$s_{t,i}^g g_{\min} \leq p_{t,i}^g \leq s_{t,i}^g g_{\max} N, \quad t \in J, i \in I \tag{4.60}$$

$$E_{t+1} = E_t + p_t^p \Delta t_h \eta_c - p_t^g \Delta t_h / \eta_d, \quad t \in K \tag{4.61}$$

$$E_{t+1,i} = E_{t,i} + p_{t,i}^p \Delta t_h \eta_c - p_{t,i}^g \Delta t_h / \eta_d, \quad t \in J, i \in I \tag{4.62}$$

$$E_{\min} \leq E_t \leq E_{\max}, \quad t \in K \tag{4.63}$$

$$E_{\min} \leq E_{t,i} \leq E_{\max}, \quad t \in J, i \in I \tag{4.64}$$

Equations (4.44) and (4.45) define the objective to maximize the revenue of the REG–ESS union. As the target power for *farther future* need not be submitted and the forecast power is not as reliable as that of the *near future*, S_t^j is multiplied by a weighting factor ω_t^p, which is smaller than 1.0. Equation (4.46) corresponds to the revenue of the future 4–8 hours (*near future*). It consists of three parts: (i) income from the net power injection to the grid, (ii) startup and shutdown costs of pumping turbines of the PHS, (iii) penalty for the deviation between day-ahead and intraday bidding (as defined in Eq. (4.50)). The latter two parts are subtracted from the revenue. Equation (4.47) corresponds to the expected revenue of the future 8–28 hours (*farther future*). As wind power scenarios are used for the *farther future*, deviation in each scenario between the output and intraday bidding is inevitable (as defined in Eq. (4.51)), which should be penalized. Penalty factors of electricity price for positive/negative deviation between actual output and biddings, i.e. γ^+ and γ^-, are set equal to the same value γ. The revenue of *farther future* is the expectation of all scenarios, which is different from that of Eq. (4.46). In the case of the *near future*, the intraday bidding is exactly the net output, as shown in Eq. (4.48).

Constraint (4.49) is a chance constraint and is equivalent to a deterministic constraint (4.65) below. The probability density function of $\xi_t(v_t)$ will be given later by (4.76).

$$0 \leq p_k^w \leq p_k^{wf} + \xi_{v_k}^{-1}(\alpha) \tag{4.65}$$

Constraint (4.52) limits the wind power fluctuation within predefined boundaries. Constraints for the on/off state of pumping and generating are given in Eq. (4.53) while the number of pumping turbines at each interval is constrained by Eqs. (4.54) and (4.55). Even though the startup and shutdown of pumping turbines are considered in the objective function with startup and shutdown costs, constraint (4.56) limits the total on–off times per day. Constraints (4.57) to (4.60) define the pumping and generating power range of the PHS, while the residue energy of PHS is constrained by Eqs. (4.61) to (4.64).

4.5.3.4 Real-Time Control and Settlement

The REG–ESS union has the following control targets in real-time operation. First of all, as the joint power output schedule was optimized in the intraday market, the union should keep the deviation between output and intraday biddings as small as

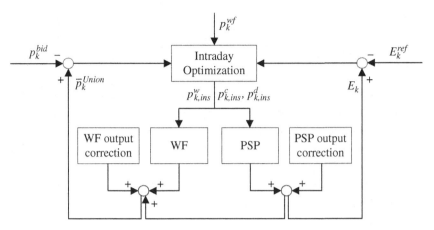

Figure 4.6 Real-time operational block diagram.

possible, in order to minimize the settlement penalty. Secondly, fluctuation of the joint output should not violate the corresponding standard. Last but not least, the residue energy of PHS should be observed and controlled in case it goes out of allowable operational range. Participation in frequency regulation and other ancillary services of an REG–ESS union is not considered in this chapter. The control scheme is shown as Figure 4.6. The blocks of *WF output correction* and *PHS output correction* represent deviations between the received instruction and the actual output of REG and ESS, which need dynamic compensation.

$$\min \left(\overline{P}_k^{union} - P_k^{bid}\right)^2 + \omega^E \left(E_k - E_k^{ref}\right)^2, \quad k \in K \tag{4.66}$$

$$\overline{P}_k^{union} = \sum_{m = \Delta t_m(k-1)+1}^{\Delta t_m k} P_m^{union}/\Delta t_m, \quad k \in K, m \in M \tag{4.67}$$

$$p_m^{union} = \begin{cases} p_{m,a}^w + p_{m,a}^g - p_{m,a}^p, & m \in M, m \leq t \\ p_{m,ins}^w + p_{m,ins}^g - p_{m,ins}^p, & m \in M, t < m \leq \Delta t_m \end{cases} \tag{4.68}$$

$$-B_f^1 \leq p_m^{union} - p_{m-1}^{union} \leq B_f^1, \quad m \in M \tag{4.69}$$

$$-B_f^{10} \leq V\left(p_m^{union}\right) \leq B_f^{10}, \quad m \in M \tag{4.70}$$

$$0 \leq p_{m,ins}^w \leq p_m^{wf}, \quad m \in M \tag{4.71}$$

$$P_{\min} n_k^p \leq p_{m,ins}^p \leq P_{\max} n_k^p, \quad k \in K, m \in M \tag{4.72}$$

$$s_k^g g_{\min} \leq p_{m,ins}^g \leq s_k^g g_{\max} N, \quad k \in K, m \in M \tag{4.73}$$

$$E_{m+1} = E_m + p_{m,ins}^p \Delta t_n \eta_c - p_{m,ins}^g \Delta t_n/\eta_d, \quad m \in M \tag{4.74}$$

$$E_{\min} \leq E_m \leq E_{\max}, \quad m \in M \tag{4.75}$$

Equation (4.66) defines the objective of real-time control, which is to bring near the intraday biddings and the average power in every 10 minutes, as well as make residue energy at the end of each interval as planned. Average power of each interval is calculated by Eq. (4.67). Constraints (4.68) define the real power output and the power instruction within a 10-minute interval. Power fluctuation within 1 minute and 10 minutes is restricted by Eqs. (4.69) and (4.70), respectively, while Eq. (4.70) can be in the form of Eqs. (4.30) or (4.31). Constraints (4.71) restrict the wind power and the pumping and generating power ranges of PHS are determined by Eqs. (4.72) and (4.73). Constraints (4.74) and (4.75) require that the residue energy of PHS should be in allowable range.

Each of the above optimization problems is a mixed integer programming problem and can be solved by off-the-shelf optimization software within an acceptable computation time.

4.5.4 Handling Wind Power Forecast Error

4.5.4.1 Description of Wind Power Forecast Error

Wind power forecast errors are usually formulated as Gaussian or beta distributions for simplicity [28]. However, it is not precise for the real cases and the forecast error distribution varies with the forecast wind speeds [29]. Wind power forecast errors obey different distributions at various forecast wind speeds. When the forecast wind speed is much lower than the cut-in wind speed or higher than the rated wind speed of the wind turbine, the forecast errors are usually small. When the forecast wind speed is near the cut-in or rated speed, the distribution shows obvious skewness and can be described as an extreme value distribution [30]. When the forecast speed is in the middle of cut-in and rated wind speed of the wind turbines, the wind forecast error can be represented by a Gaussian distribution. A versatile probability model of wind power forecast errors was put forward in reference [31], which can be easily applied in the chance constrained optimization. The distribution of wind power forecast errors is described as a classified versatile probability function, which can reveal detailed information and convert chance constraints to deterministic ones. The versatile probability density function proposed in reference [31] is

$$f(x|\nu,\lambda,\mu) = \frac{\nu\lambda e^{-\nu(x-\mu)}}{\left(1 + e^{-\nu(x-\mu)}\right)^{\lambda+1}} \tag{4.76}$$

The cumulative density function and its inverse function are as follows:

$$F(x|\nu,\lambda,\mu) = \left(1 + e^{-\nu(x-\mu)}\right)^{-\lambda} \tag{4.77}$$

$$F^{-1}(c|\nu,\lambda,\mu) = \mu - \frac{1}{\nu}\ln\left(c^{-1/\lambda} - 1\right) \tag{4.78}$$

The distribution given by Eq. (4.76) has two main advantages over commonly used distributions such as Gaussian and beta distributions. Firstly, it can represent different kinds of distribution with various parameters. Secondly, it has an explicit inverse function, which makes it easier to transfer the chance constraints to deterministic ones. Parameters of the function can be fitted for every wind speed interval, e.g. 1 m s^{-1}, in order to model different wind power forecast errors more precisely at different speeds.

4.5.4.2 Determination of the Weighting Factor

The forecast errors of wind power are affected not only by forecast wind speeds but also by the time horizon (lead time). The influence of lead time on forecast error was studied in reference [7], which shows that the forecast deviation increases linearly or weakly quadratically with lead time. Here, it is assumed that this deviation increases linearly with lead time. Therefore, the forecast error related to wind speeds should be multiplied by the per-unit of forecast error variance proportional to the lead time, to determine the final forecast error, as shown in Figure 4.7.

Weighting factors $\{\omega_t\}$ play an important role in intraday optimization. As mentioned above, ω_t reflects the influence of the forecast error and there is a set of weighting factors that should be decided by the forecast errors [32]. For the forecast error set $\varepsilon = \{e_1, e_2, ..., e_n\}$, the corresponding weighting factor is defined as follows:

$$\omega_t = \exp \left\{ -\frac{e_t - \min\{e_t\}}{\max\{e_t\} - \min\{e_t\}} \right\} \tag{4.79}$$

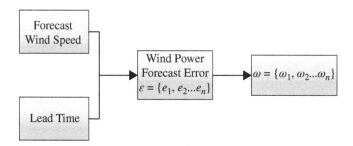

Figure 4.7 Procedure for determining the weighting factor for farther future periods.

4.5.5 Case Studies

A simple case is designed to study the revenue enhancement brought by the intra-day market and rolling optimization strategy. Bidding results of rolling optimization are given to show that the detailed procedure and sensitivity analyses on key parameters of the model are carried out.

The REG–ESS union consists of a 495-MW WF group, which is in the northeast of China, and a 125-MW PHS plant. The PHS plant consists of five identical reversible pumping/generating turbines, whose basic parameters are given in Table 4.3. The startup cost is equal to the shutdown cost for each unit, which is set to 1000 RMB per event [4]. Electricity prices are forecasted through historical data and it is assumed that the forecast price is accurate. In some countries, the electricity price is determined by local government and varies in different regions. In the following simulations, the electricity price within the period of 9:00 to 23:00 is 0.8 RMB kWh^{-1} and the price of the rest period is 0.44 RMB kWh^{-1}, which is same as that of the case study in Section 3.3.3 of Chapter 3. As electricity prices are forecasted or given in advance, they are treated as constants in the optimization formulation.

The electricity prices for a PHS to generate and consume energy are usually different if the PHS is not bidding in a power market. Where there is no competing wholesale electricity market, different PHS plants may have different power generation and consumption prices to recover their investment and operation costs. In the case of the following simulations, the optimal operation of PHS under a deregulated power market environment is assumed. The wholesale prices of electricity generation and consumption are the same for a same place in the power market. The price variations will incentivize optimal operation of the PHS and other types of ESS to harvest more revenue through price arbitrage.

The penalty factor γ for deviation between output and intraday bidding should be determined based on the balancing cost paid to other market participants, and is set equal to 0.44 here. The penalty to deviation between day-ahead and intraday biddings should represent the cost of rescheduling other generators or controllable resources. However, as this penalty is not implemented in the actual power market, the factor β is set at a moderate value, 0.22, in the case studies, and results of the sensitivity analysis will be given later.

Table 4.3 Basic parameters of PHS.

E_{min} (MWh)	E_{max} (MWh)	p_{min} (MW)	p_{max} (MW)	g_{min} (MW)	g_{max} (MW)	η_c, η_d
50	500	18	22	10	25	0.9

4.5.5.1 Bidding Results

According to the rolling optimization mechanism, the day-ahead bidding occurs at 12:00 of Day $(D-1)$ and six intraday biddings occur every four hours in Day D. The biddings, control strategies, and settlements are all for Day D. The day-ahead bidding is illustrated in Figure 4.8 as it is the premise of intraday bidding. The thin curves represent six distinct wind power scenarios that are obtained by simulation based on the forecast wind power and the forecast error distributions.

It can be seen from Figure 4.8 that the PHS pumps water storing energy when the electricity price is low and generates it when the price is high. Consequently, in the price valley period from 23:00 to 9:00, the bidding power is less than the average forecast wind power and vice versa for price peak periods.

Figure 4.9 compares the day-ahead and intraday biddings. With a more accurate wind power forecast, the REG–ESS union would revise its day-ahead bidding in the intraday market when the forecast difference exceeds by some extent. By periodically updating bidding in the intraday market, the union can reduce the deviation between the final bidding and its actual total power output. In this way, the union can mitigate the penalty for output deviation and increase the total revenue.

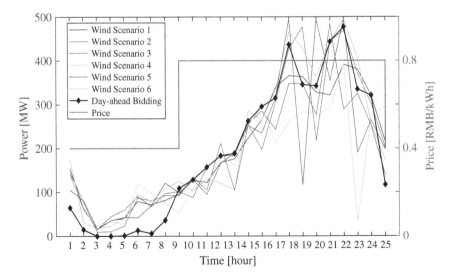

Figure 4.8 Day-ahead optimal bidding of the REG–ESS union.

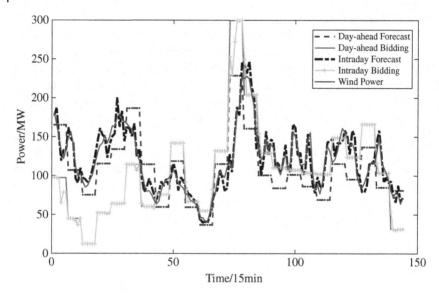

Figure 4.9 Comparison of day-ahead and intraday biddings.

4.5.5.2 Revenue Comparisons

On the one hand, with the help of intraday markets, the union can revise its biddings at a relatively low cost, which saves at least part of the penalty for output deviation. On the other hand, PHS adjusts its output schedule according to an updated wind power forecast.

Take a specific day, for example. Table 4.4 compares all the items that influence the total revenue of the union. When there are intraday markets, the deviation penalty decreases dramatically by over 50%, which accounts for a major part of the total revenue improvement. The reduction in penalty comes at the cost of the intraday rebidding cost, which is relatively small because of the lower penalty factor β. Comparing this to the case without intraday markets, the income from selling electricity drops by 1.5% with the startup/shutdown cost unchanged. The total revenue of the REG–ESS union increases by 4.0% with rolling optimization.

If running separately, the wind farm participates in the energy market alone and makes a revenue through rolling optimization, while the PHS makes a revenue through price arbitrage. The total revenue of the two entities is 1 794 993 RMB and is lower than the revenue of a joint operation by 3.70%, which proves the effectiveness of coordinated operation. Details of the revenue comparisons are listed in Table 4.5.

It can be calculated from Table 4.5 that the wind farm increases revenue by 103 301 RMB through coordination while the PHS's revenue decreases by 34 335 RMB.

Table 4.4 Comparison of revenue with/without intraday markets (RMB).

Items	Selling energy revenue	Startup/ shutdown cost	Deviation penalty	Rebidding cost	Total revenue
Without intraday market	1 999 446	7000	200 233	/	1 792 213
With intraday market	1 968 298	7000	87 357	9982	1 863 959

Table 4.5 Revenue comparisons with/without coordination (RMB).

Coordination	REG	ESS	Sum	Increase rate
W/O	1 668 593	126 400	1 794 993	/
W	1 771 894	92 065	1 863 959	3.84%

In order to ensure that both participants can benefit from the coordination, the wind farm should share partial increased revenue with ESS. The transferred revenue can be in the range from 34 335 to 103 301 RMB theoretically, but the actual subsidy will depend on the contract between them.

4.5.5.3 Sensitivity Analysis

In order to study the influence that some parameters, such as reservoir volume, installed capacity, and penalty factors, have on the total revenue, sensitivity analyses are carried out based on historical data. For simplicity of the following discussion, we define *Scenario 1* as the case with a day-ahead market only and *Scenario 2* as the case with rolling optimization of a complete set of day-ahead/ intraday biddings and real-time operation. In the sensitivity analysis, the revenues of *Scenario 1* and *Scenarios 2* are compared at the same time.

Penalty factor β for deviation between intraday and day-ahead biddings is essential to the intraday rebidding section. Figure 4.10 demonstrates its influence on the revenue improvement by intraday rebidding. For each value of β, 50 scenarios of wind power are simulated and the mean value is represented by the blue dots. When the factor is zero, the revenue improvement is the value of updated wind power forecasts. When the penalty factor β is over 0.34, the revenue of intraday bidding is even less than that of the day-ahead market. This means that only if $\beta < 0.34$, the REG–ESS union would get more revenue through the intraday market. The turning point, $\beta = 0.34$, means that 34% of the wind power deviation would be penalized for rescheduling other generators.

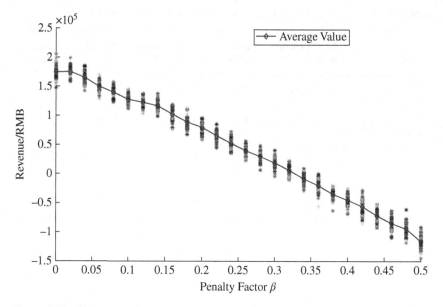

Figure 4.10 Effects of penalty factor β on the improved revenue by the intraday market.

Regarding the penalty factors of an output deviation, it is clear from Figure 4.11 that with the increase of the positive/negative penalty factor, the revenues of both *Scenarios 1* and *2* drop gradually. *Scenario 2* is always occupying an advantage over *Scenario 1*, which demonstrates the meaning of rolling optimization in promoting revenue of the WF–PHS union. Besides, the influence of the positive factor on the total revenue is more prominent than the negative factor. The phenomenon can be explained as follows. When the negative factor increases, the output deviation penalty increases but it also gives the union an incentive to generate more, which partially compensates for the penalty. However, the growth of the positive penalty factor, on the one hand, raises the deviation penalty and, on the other hand, gives the union an incentive to reduce their output, which worsens the overall revenue. Consequently, revenue of the WF–PHS union is more sensitive to the positive penalty factor than the negative one.

Figure 4.12 shows the change of the total revenue of the REG–ESS union with the overall installed capacity of the PHS, namely the number of installed pumping/generating turbines. Revenue with intraday markets will keep steady if there are over six turbines, mainly because there is already enough power capability and other factors such as wind power forecast accuracy constrain further improvement of revenue. The phenomenon is more obvious for the scenario with rolling optimization as the inflection point is three. The comparison also proves that the rolling optimization can better harvest the potential value of ESS.

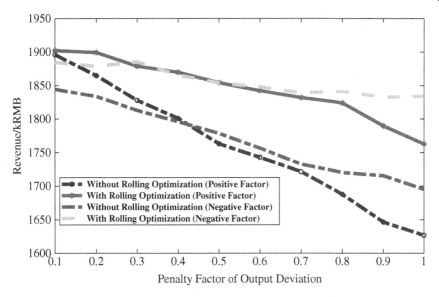

Figure 4.11 Effects of output deviation penalty factors γ^+ and γ^- on revenue.

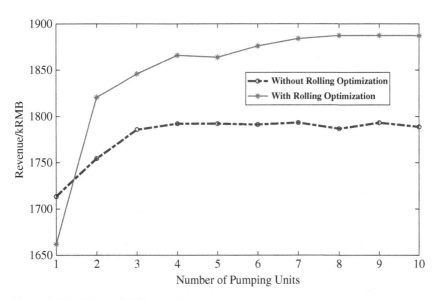

Figure 4.12 Effects of PHS capacity on revenue.

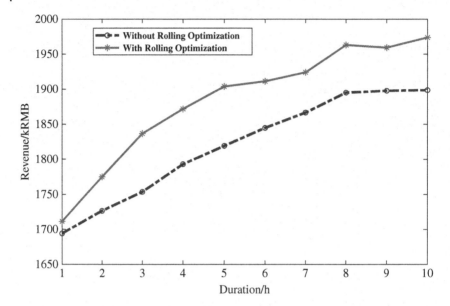

Figure 4.13 Effects of reservoir volume of PHS on revenue.

When the power capacity of PHS is fixed, the duration of pumping/generating at the rated power is equivalent to the reservoir volume of PHS. Figure 4.13 indicates that with the expansion of the reservoir, revenue increases monotonously. When the duration reaches 8 hours, revenue will rise slowly, because the bottleneck ascribes to other factors, such as the overall capacity and wind power forecast accuracy.

4.6 Conclusion and Discussion

In the first part of this chapter, an integrated strategy for an REG–ESS union is derived to optimally determine offers as a price taker in day-ahead markets and operation policy at the balancing stage, based on linear decision rules. The strategy is modeled as a stochastic optimization problem taking uncertainties of market prices and wind power generation into account. The concept of energy value is additionally introduced to consider the value of the residual energy of ESS in the objective function. The objective function controls the balance between expected revenue and risk aversion. The advantage of this strategy will be more prominent in the case of higher uncertainty and dynamic prices. The REG–ESS union could tune its strategy based on risk preferences and then make a trade-off between the expected revenue and risk. The second part of this chapter discusses a modified market design that allows for day-ahead bidding and intraday rebidding, which can utilize the updated REG power forecast information better. The penalty for deviation between day-ahead and intraday biddings is optimally set, which on the one hand can encourage REGs to improve their forecast accuracy and on the other hand can motivate the participants to revise their biddings in the intraday market with a relatively low deviation penalty. The setup of the intraday market not only improves the revenue of electricity market participants but also helps to reduce the dispatch adjustments for the system operator. According to the discussed market mechanism, stochastic optimization models are formulated and solved to maximize the joint revenue of an REG–ESS union. In an intraday market optimization, the chance-constrained formulation is transferred to a deterministic one with the help of a versatile probability function, which is used to describe the wind power forecast error and has an explicit inverse form. As the forecast error increases with the time horizon, the time periods of *near future* (4–8 hours) and *farther future* (8–28 hours) are dealt with separately and differently. Case studies show that with the help of a rolling optimization, an REG–ESS union can fully utilize the energy storage capability of the ESS and increase the total revenue. Being allowed to revise bidding in the intraday markets, an REG–ESS union can take advantage of updated forecast information and further enhance the overall revenue.

References

1 Taylor, J., Callaway, D.S., and Poolla, K. (May 2013). Competitive energy storage in the presence of renewables. *IEEE Transactions on Power Apparatus and Systems* 28 (2): 985–996.

2 Castronuovo, E.D., Usaola, J., Bessa, R. et al. (June 2014). An integrated approach for optimal coordination of wind power and hydro pumping storage. *Wind Energy* 17 (6): 829–852.

3 Chen, X., Sim, M., Sun, P., and Zhang, J. (2008). A linear decision-based approximation approach to stochastic programming. *Operations Research* 56 (2): 344–357.

4 Ding, H., Pinson, P., Hu, Z., and Song, Y. (November 2016). Optimal offering and operating strategies for wind-storage systems with linear decision rules. *IEEE Transactions on Power Systems* 31 (6): 4755–4764.

5 Tuohy, A., Denny, E., and O'Malley, M. (2007). Rolling unit commitment for systems with significant installed wind capacity. *Power Tech, 2007*, Lausanne. IEEE, pp. 1380–1385.

6 Jafari, A.M., Zareipour, H., Schellenberg, A., and Amjady, N. (May. 2014). The value of intra-day markets in power systems with high wind power penetration. *IEEE Transactions on Power Systems* 29: 1–12.

7 Menemenlis, N., Huneault, M., and Robitaille, A. (October 2012). Computation of dynamic operating balancing reserve for wind power integration for the time-horizon 1–48 hours. *IEEE Transactions on Power Systems* 3: 692–702.

8 Pinson, P., Chevallier, C., and Kariniotakis, G.N. (August 2007). Trading wind generation from short-term probabilistic forecasts of wind power. *IEEE Transactions on Power Systems* 22: 1148–1156.

9 Morthorst, P.E. (2007). Further developing Europe's power market for large scale integration of wind power: D4. 1. Detailed investigation of electricity market rules. Cases for France, Germany, Netherlands, Spain and Denmark. European Wind Energy Association (EWEA).

10 Grønvik, I., Hadziomerovic, A., Ingvoldstad, N. et al. (2014). Feasibility of linear decision rules for hydropower scheduling. *Proceedings of the 2014 IEEE International Conference on Probabilistic Methods Applied to Power Systystems*. IEEE, pp. 1–6.

11 Rockafellar, R.T. and Uryasev, S. (2002). Conditional value-at-risk for general loss distributions. *Journal of Banking and Finance* 26 (7): 1443–1471.

12 Ding, H., Pinson, P., Hu, Z., and Song, Y. (January 2016). Integrated bidding and operating strategies for wind-storage systems. *IEEE Transactions on Sustainable Energy* 7 (1): 163–172.

13 Morales, J.M., Conejo, A.J., Madsen, H. et al. (2014). *Integrating Renewables in Electricity Markets – Operational Problems*. New York, NY, USA: Springer.

14 Rockafellar, R.T. and Uryasev, S. (2000). Optimization of conditional value-at-risk. *Journal of Risk* 2: 21–42.

15 Krokmahl, P., Palmquist, J., and Uryasev, S. (2002). Portfolio optimization with conditional value at risk objective and constraints. *Journal of Risk* 4: 43–68.

16 Uryasev, S. (2000). Conditional value-at-risk: Optimization algorithms and applications. *Financial Engineering News* 14: 3–8.

17 Warrington, J., Goulart, P., Mariéthoz, S., and Morari, M. (November 2013). Policy-based reserves for power systems. *IEEE Transactions on Power Apparatus and Systems* 28 (4): 4427–4437.

18 Energinet, download of market data [Online]. http://energinet.dk/EN/El/Engrosmarked/Udtraek-af-markedsdata/Sider/default.aspx.

19 Pinson, P. (November 2013). Wind energy: Forecasting challenges for its operational management. *Statistical Science* 28 (4): 564–585.

20 Bukhsh, W.A., Zhang, C., and Pinson, P. (May 2016). An Integrated Multiperiod OPF Model with Demand Response and Renewable Generation Uncertainty. *IEEE Transactions on Smart Grid* 7 (3): 1495–1503.

21 Bukhsh, W.A. (2014). Data for stochastic multiperiod optimal power flow problem [Online]. https://sites.google.com/site/datasmopf/intro.

22 Aasgård, E.K., Andersen, G.S., Fleten, S.E., and Haugstvedt, D. (July 2014). Evaluating a stochastic-programming-based bidding model for a multireservoir system. *IEEE Transactions on Power Apparatus and Systems* 29 (4): 1748–1757.

23 Dent, C.J., Bialek, J.W., and Hobbs, B.F. (August 2011). Opportunity cost bidding by wind generators in forward markets: Analytical results. *IEEE Transactions on Power Apparatus and Systems* 26 (3): 1600–1608.

24 Castronuovo, E.D., Usaola, J., Bessa, R. et al. (June 2014). An integrated approach for optimal coordination of wind power and hydro pumping storage. *Wind Energy* 17 (6): 829–852.

25 Bourry, F., Costa, L.M., and Kariniotakis, G. (2009). Risk-based strategies for wind/pumped-hydro coordination under electricity markets. *Power Tech, 2009.* IEEE, Bucharest, pp. 1–8.

26 Garcia-Gonzalez, J., de la Muela, R.M.R., Santos, L.M., and Gonzalez, A.M. (May 2008). Stochastic joint optimization of wind generation and pumped-storage units in an electricity market. *IEEE Transactions on Power Systems* 23: 460–468.

27 Fox, B. (2007). *Wind Power Integration: Connection and System Operational Aspects*, 50e. Institution of Engineering and Technology.

28 Bludszuweit, H., Dominguez-Navarro, J.A., and Llombart, A. (August 2008). Statistical Analysis of Wind Power Forecast Error. *IEEE Transactions on Power Systems* 23 (3): 983–991.

29 Hodge, B. and Milligan, M. (2011). Wind power forecasting error distributions over multiple timescales. *2011 IEEE Power and Energy Society General Meeting*, Detroit, MI, USA, pp. 1–8.

30 Ganger, D., Zhang, J., and Vittal, V. (November 2014). Statistical characterization of wind power ramps via extreme value analysis. *IEEE Transactions on Power Systems* 29 (6): 3118–3119.

31 Zhang, Z., Sun, Y., Gao, D.W. et al. (August 2013). A versatile probability distribution model for wind power forecast errors and its application in economic dispatch. *IEEE Transactions on Power Systems* 28 (3): 3114–3125.

32 Tewari, S., Geyer, C.J., and Mohan, N. (November 2011). A statistical model for wind power forecast error and its application to the estimation of penalties in liberalized markets. *IEEE Transactions on Power Systems* 26: 2031–2039.

5

Unit Commitment with Energy Storage System

5.1 Introduction

Because load changes with time, the generating units should be properly committed in order to serve the load with the least cost while satisfying specified constraints. The constraints include unit technical constraints, system security, or reliability constraints, and so forth, depending on the generation mix, load–curve characteristics, and so on [1]. For a large-scale power system with hundreds of units, the unit commitment (UC) problem is a complex large scale optimization problem and very difficult to solve. This challenging problem has attracted many researchers for decades and many literatures have been published [2–4].

Traditionally, the most common methods are priority list, dynamic programming, and Lagrange relaxation. A lot of modern heuristic optimization methods have been used to solve the UC problem, including genetic algorithm, simulated annealing, and tabu search, because they are easy to implement and powerful in searching for a global optimum [3]. With the development of mixed-integer linear programming (MILP) algorithm, off-the-shelf optimization software, and enhanced computation power, MILP-based methods are now widely used [5]. With the third edition of *Power Generation, Operation, and Control* [1], MILP is added as one of the most talked-about techniques for the solution of the UC problem. Another important factor that promotes the application of MILP is the development of power markets. In reference [6], the UC model and solution method required under the deregulated environment are discussed in detail.

In recent years, with the increasing penetration of renewable energy generation (REG), the focus on UC is transitioning from traditional deterministic approaches to stochastic optimization ones. To deal with the uncertain wind and/or solar power output, various formulations and solution approaches have been applied to UC. The three main types of them are stochastic programming, robust optimization,

Energy Storage for Power System Planning and Operation, First Edition. Zechun Hu.
© 2020 John Wiley & Sons Singapore Pte. Ltd.
Published 2020 by John Wiley & Sons Singapore Pte. Ltd.

and stochastic dynamic programming [4]. The challenges include uncertainty modeling, computational requirements, and the balance between low cost and high reliability of power system operation.

There are a number of published papers on UC with large-scale energy storage system [7–12]. In reference [7], a method for UC with ESS and thermal units is proposed, based on the extended priority list method. Both wind power and ESS are considered in the UC problem in references [8] and [9]. However, the uncertainty of wind power is not dealt with in reference [8], while a robust optimization approach to accommodate wind output uncertainty is proposed in reference [9] to obtain the least cost schedule under the worst wind power output scenario. In reference [10], the authors present a mathematical definition of an ideal and generic storage device. A *generic storage device* is defined as any device with the capability of transforming and storing energy, and reverting the process by injecting back the stored energy to the system. Furthermore, an *ideal storage device* assumes certain simplifications in its technical and economic operation. Under the assumptions for the generic and ideal ESS, models of ESS for both deterministic and scenario-based stochastic UCs are built. The fast responsive capability of utility-scale battery energy storage is exploited in the security-constrained unit commitment (SCUC) problem [11]. Fast corrective control under severe contingencies is considered using battery ESS. In reference [12], pumped hydroelectric storage is considered to balance the mismatches between demand and generation including wind power. An interval UC formulation is built to take the forecast errors into account with the new PHS constraints. It should be noted that network constraints are omitted.

In this chapter, we will focus on the SCUC formulations considering large-scale ESS. Only thermal generating units are considered in the SCUC problem. The objective is to minimize the total operational cost under a centrally dispatching model, which can be easily extended to maximize the social benefit under a deregulated environment.

5.2 Energy Storage System Model for SCUC

In Chapter 2, the steady-state models of three types of energy storage system have been derived. To make it easier to understand, a general model of BESS for steady-state operation is given as follows.

1) Energy transition constraint

It is assumed there are M BESSs deployed in a power system. For each BESS:

$$E_{m,t} = E_{m,t-1} + \left(p_{em,t}^c \times \eta_{m,c} - p_{em,t}^d / \eta_{m,d}\right) \times \Delta t \tag{5.1}$$

where $t \in \psi_t = \{1, 2, \ldots, T\}$ and $m \in \psi_{es}$. They have the same meanings in (5.2)–(5.7) below. $E_{m,t}$ is the residue energy stored by BESS m at time t and $E_{m,0}$ is the given initial value; $p^c_{em,t}$ and $p^d_{em,t}$ are the charging and discharging powers of BESS m at time t, respectively; and $\eta_{m,c}$ and $\eta_{m,d}$ stand for the charging and discharging efficiencies of BESS m, respectively.

2) Charging/discharging power limits

$$u^c_{em,t} \times p^{c\,\min}_{em} \leq p^c_{em,t} \leq u^c_{em,t} \times p^{c\,\max}_{em} \tag{5.2}$$

where $u^c_{em,t}$ represents the charging state of BESS m at time t, with 1 for charging, 0 for being idle, and $p^{c\,\min}_{em}$ and $p^{c\,\max}_{em}$ are the minimum and maximum charging powers of BESS m, respectively.

$$u^d_{em,t} \times p^{d\,\min}_{em} \leq p^d_{em,t} \leq u^d_{em,t} \times p^{d\,\max}_{em} \tag{5.3}$$

where $u^d_{em,t}$ represents the discharging state of BESS m at time t, with 1 for discharging and 0 for being idle, and $p^{d\,\min}_{em}$ and $p^{d\,\max}_{em}$ are the minimum and maximum discharging powers of BESS m, respectively.

3) Stored energy limits

For chemical batteries, it is common for the capacity degradation to decrease with the reduced depth of discharge [13]. Thus, the following limits are set up to improve its cycle life:

$$E^{\min}_m \leq E_{m,t} \leq E^{\max}_m \tag{5.4}$$

where $E^{\min}_m$ and $E^{\max}_m$ are the minimum and maximum energy storage levels of BESS m, respectively.

4) Charging/discharging states limit

In each time interval, a BESS can only be in a charging, discharging, or idle state, which is constrained by the following constraint:

$$0 \leq u^c_{em,t} + u^d_{em,t} \leq 1 \tag{5.5}$$

5) Reserve capacity limits

$$0 \leq p^r_{em,t} + p^d_{em,t} - p^c_{em,t} \leq p^{d\,\max}_{em} \tag{5.6}$$

$$E^{\min}_m \leq E_{m,t} - \left(p^r_{em,t} + p^d_{em,t} - p^c_{em,t}\right)/\eta_d \times \Delta t \tag{5.7}$$

Constraint (5.6) ensures that: (i) a BESS can increase discharging power for reserve support when it is in the discharging or idle mode; (ii) when in a charging state, a BESS can be treated as a controllable load to reduce charging power and change the operation mode for providing reserve capacity, and (iii) the reserve capacity and the discharging power together should not be greater than the maximum discharging power.

This general model (5.1) to (5.7) for steady-state operation of BESS takes into account the charging and discharging efficiencies and available reserve capacity of BESS.

5.3 Deterministic SCUC with Energy Storage System

In order to supply the forecast load with a minimum production cost, an SCUC model is formulated to optimally dispatch both thermal generation units and BESSs.

5.3.1 Objective Function

The objective function is to minimize the total cost including the generation and startup costs (SCs) of thermal plants, as shown in the following formula:

$$\min \sum_{t \in \psi_t} \sum_{i \in \psi_g} \left[u_{i,t} F_i(p_{i,t}) + S_i u_{i,t}(1 - u_{i,t-1}) \right] \tag{5.8}$$

where ψ_t and ψ_g are the sets of time interval and generator, respectively, $u_{i,t}$ denotes the state of unit i at time t with 1 for on and 0 for off, S_i is the SC function of unit i, and $F_i(\cdot)$ is the production cost function of unit i. In order to solve the SCUC model efficiently, the quadratic production cost function $F_i(p_{i,t})$ can be approximated using the piecewise linear approximation method [14]. It should be noted that the shutdown costs of the units are not considered in the objective function. Furthermore, the operation cost (OC) of BESS can also be added to the objective function when counting the degradation cost of the battery.

5.3.2 Constraints

The constraints of SCUC include power flow constraints, generating unit technical constraints, and power network security constraints. All the constraints considered are listed below.

1) Power balance constraint
   The DC power flow is considered in this formulation and the total active power balance constraint of the system is as follows:

$$\sum_{i=1}^{N} p_{i,t} - \sum_{m=1}^{M} \left(p_{em,t}^c - p_{em,t}^d \right) = P_{d,t}, \forall t \in \psi_t \tag{5.9}$$

where $P_{d,t}$ is the total active power demand at time t.

2) System operating reserve requirement

$$\sum_{i=1}^{N} u_{i,t} P_i^{\max} - \sum_{m=1}^{M} \left(p_{em,t}^c - p_{em,t}^d - p_{em,t}^r \right) \geq (1 + \gamma) P_{d,t}, \forall t \in \psi_t \tag{5.10}$$

where γ is the required spinning reserve rate. It should be noted that the reserve support from the BESSs is considered in the above constraints.

3) Generating unit constraints

For the thermal units, the unit operating constraints include minimum and maximum power outputs, ramping up/down limits, and minimum up/down time limits [15].

• Power output limits

$$u_{i,t} \times P_i^{\min} \leq p_{i,t} \leq u_{i,t} \times P_i^{\max}, \forall i \in \psi_g, t \in \psi_t \tag{5.11}$$

where i denotes the ith generation unit, $P_i^{\min}$ and $P_i^{\max}$ are the minimum and maximum power outputs of unit i, respectively, and $p_{i,t}$ is the power output of unit i at time t.

• Ramping up/down limits [14]

$$p_{i,t} - p_{i,t-1} \leq u_{i,t-1} \times RU_i + P_i^{\max} \times (1 - u_{i,t-1}), \forall i \in \psi_g, t \in \psi_t \tag{5.12}$$

$$p_{i,t} - p_{i,t-1} \geq -u_{i,t-1} \times RD_i - P_i^{\max} \times (1 - u_{i,t}), \forall i \in \psi_g, t \in \psi_t \tag{5.13}$$

where $RD_{i,t}$ and $RU_{i,t}$ are the ramp-down and ramp-up limits of unit i within a time interval, respectively.

• Minimum up/down time limits

$$\sum_{k=t-SU_i+1}^{t} su_{i,k} \leq u_{i,t}, \forall i \in \psi_g, t \in \psi_t \tag{5.14}$$

$$\sum_{k=t-SD_i+1}^{t} sd_{i,k} \leq 1 - u_{i,t}, \forall i \in \psi_g, t \in \psi_t \tag{5.15}$$

where $su_{i,t}$ and $sd_{i,t}$ are the startup and shutdown binary flags of unit i at time t (1 for startup or shutdown and 0 for no action) and SU_i and SD_i are the minimum startup and shutdown times of unit i, respectively.

4) State switching limits of BESSs

For a BESS, the total times of state switching may be capped:

$$\sum_{t=1}^{T} \left(\left| u_{em,t}^c - u_{em,t-1}^c \right| + \left| u_{em,t}^d - u_{em,t-1}^d \right| \right) \leq k_m^{\max}, \forall m \in \psi_{es} \tag{5.16}$$

where $k_m^{\max}$ is the preset maximum number of state switchings.

5) Constraints on BESSs
   For each BESS, Eqs. (5.1) to (5.7) should be included.
6) Allowable residue energy difference within a dispatch cycle
   The difference between the residual energy stored in a BESS at the starting and ending times of a dispatch cycle, e.g. a day, is confined:

$$|E_{m,0} - E_{m,T}| \leq \varepsilon E_{m,0}, \forall m \in \psi_{es} \tag{5.17}$$

where ε is the predefined maximum allowable percentage.

7) DC power flow limits
   To securely operate a power system, the power flows from each branch should not violate its limit. For the DC flow model, the branch flow can be calculated and restricted as follows:

$$-F_l^{\max} \leq \sum_{i \in \psi_n} K_{l,i} \left(\sum_{j \in \psi_{gi}} P_{j,t} - P_{di,t} + P_{ei,t}^d - P_{ei,t}^c \right) \leq F_l^{\max}, \forall l \in \psi_l \tag{5.18}$$

where ψ_n, ψ_{gi}, and ψ_l are the sets of nodes, generating units at node i and lines, respectively, $F_l^{\max}$ is the maximum power limit of line l, and $K_{l,i}$ is the line flow distribution factor for the line l to the net injection at bus i [1].

By linearizing the generation cost function of each unit, the SCUC formulation given above can be formed as an MILP problem. Then the problem can be solved with optimization software package such as CPLEX.

5.3.3 Case Studies

5.3.3.1 Test System
The IEEE RTS 24-bus system with 26 generators [16] is selected as the test system. This system is divided into two areas according to the voltage levels (as shown in Figure 5.1). Only the inter-area transmission capacity is considered, which is limited to 800 MW. The operating reserve rate is set equal to 5% of the load demand.

Five BESSs are added to the test system and each BESS has a rated power of 40 MW and a storage capacity of 200 MWh. These BESSs are connected to the system at five nodes, i.e. nodes 3, 4, 5, 6, 8. The charging and discharging efficiencies of each BESS are set equal to 0.95 and 0.92, respectively. The initial stored energy in each BESS is assumed to be 30% of the rated capacity and the maximum allowable variation of residue energy is set to ±5% of the total energy capacity at the end of the day.

5.3.3.2 Simulation Scenarios
The following three different scenarios are simulated and compared:

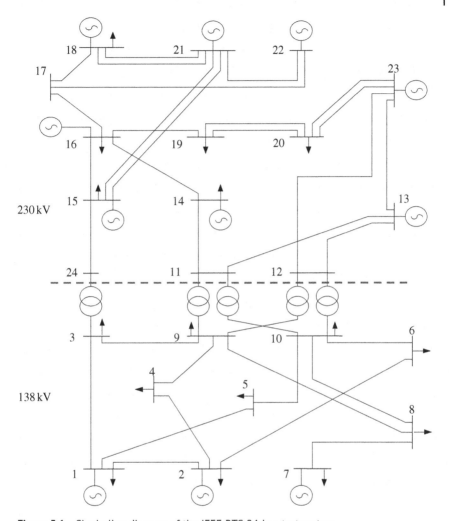

Figure 5.1 Single line diagram of the IEEE RTS 24-bus test system.

1) Scenario A – the BESSs are operated for both peak-load shaving and reserve support.
2) Scenario B – the BESSs are operated only for peak-load shaving.
3) Scenario C – all the BESSs are out of service and this is actually the base case.

Simulation results are compared from the following four perspectives:

1) Total production cost.
2) Equivalent load profile (load power minus BESS power output).

Table 5.1 Production costs for the three simulated scenarios.

Scenario	SC ($)	OC ($)	Total production cost ($)
A	2900	747, 856.22	750, 756.22
B	3090	750, 536.82	753, 626.82
C	3240	758, 684.73	761, 924.73

3) Contribution of BESSs for providing spinning reserve.
4) Influences of different sizes of BESS capacities on the SCUC results.

5.3.3.3 Simulation Results

Table 5.1 lists the total production costs including the SC and OC under the three scenarios. It can be seen from Table 5.1 that the total production cost under Scenario A is lower than the ones under Scenarios B and C by $11 168.51 (1.49% of the total cost) and $2870.60 (0.38% of the total cost), respectively. This can be explained by the following facts:

1) The energy flow in a BESS is bidirectional and can inject power to the network as an energy source when the load is high so that more expensive generation units do not need to be committed.
2) If the residual power of a BESS (the maximum discharging power minus the actual discharging power) are non-zero, it can be used for spinning reserve. BESSs can release some traditional generation units from the duty of providing spinning reserve so that these generating units can be operated more efficiently to achieve more economic dispatch.
3) With BESSs, the SC is decreased because the startup times of some units are reduced.
4) Using BESSs for reserve support, some generation units can stay uncommitted, e.g. Unit 23 in Table 5.2, so that the production cost is further reduced.

Figure 5.2 depicts the output power of traditional units in the three scenarios. With BESSs, the total output of the traditional units is reduced during peak hours and increased during the valley period. When using BESSs for both peak-load shaving and reserve support (Scenario A), the unit output can be further flattened compared to Scenario B.

Different from the one in Scenario B, the spinning reserve in Scenario A is provided by the generation units, i.e. $\sum_{i=1}^{N} u_{i,t}\left(P_i^{\max} - P_{i,t}\right)$, as well as BESSs, i.e. $\sum_{m=1}^{M} p_{em,t}^r$. The combination of the spinning reserve obtained based on

Table 5.2 Scheduling results of Scenario A and B.

Unit number	1	2	3	4	5	6	7	8	9	10	11	12	13	14	15	16	17	18	19	20	21	22	23	24
U1	1	0	0	0	0	0	0	0	0/1	0/1	0/1	0/1	0/1	0/1	0/1	0/1	0	1/0	1/0	0	0	0	0/1	1
U2	1	0	0	0	0	0	0	0	0/1	0/1	0/1	0/1	0/1	0/1	0/1	0/1	0	1/0	1/0	0	0	0	0/1	1
U3	1	0	0	0	0	0	0	0	0/1	0/1	0/1	0/1	0/1	0/1	0/1	0/1	0	1/0	0	0	0	0	0/1	1
U4	1	0	0	0	0	0	0	0	0/1	0/1	0/1	0/1	0/1	0/1	0/1	0/1	0	1/0	0	0	0	0	0	1
U5	1	0	0	0	0	0	0	0	0	0/1	0	0/1	0	0/1	0	0/1	0	1/0	1/0	0	0	1/0	0	1
U6	0/1	0	0	0	0	0	0	0	0	0	0	0	0	0	0	0	0	0	0	0	0	0	0	0
U7	0	0	0	0	0	0	0	0	0	0	0	0	0	0	0	0	0	0	0	0	0	0	0	0
U8	0/1	0	0	0	0	0	0	0	0	0	0	0	0	0	0	0	0	0	0	0	0	0	0	0
U9	0	0	0	0	0	0	0	0	0	0	0	0	0	0	0	0	0	0	0	0	0	0	0	0
U10	1	1	1	1	1	1	1	1	1	1	1	1	1	1	1	1	1	1	1	1	1	1	1	1
U11	1	1	1	1	0/1	0/1	0/1	1	1	1	1	1	1	1	1	1	1	1	1	1	1	1	1	1
U12	1	1	1	1	1	1	1	1	1	1	1	1	1	1	1	1	1	1	1	1	1	1	1	1
U13	1	1	1	1	1	1	1	1	1	1	1	1	1	1	1	1	1	1	1	1	1	1	1	1
U14	1	1	1	1	0	0	0	0/1	1	1	1	1	1	1	1	1	1	1	1	1	1	1	1	1
U15	1	1	1	1	0	0	0	0	1	1	1	1	1	1	1	1	1	1	1	1	1	1	1	1
U16	0	0	0	0	0	0	0	0	0	1	1	1	1	1	1	1	1	1	1	1	1	1	1	1
U17	1	1	1	1	1	1	1	1	1	1	1	1	1	1	1	1	1	1	1	1	1	1	1	1
U18	1	1	1	1	1	1	1	1	1	1	1	1	1	1	1	1	1	1	1	1	1	1	1	1

(Continued)

Table 5.2 (Continued)

Unit number	1	2	3	4	5	6	7	8	9	10	11	12	13	14	15	16	17	18	19	20	21	22	23	24
U19	1	1	1	1	1	1	1	1	1	1	1	1	1	1	1	1	1	1	1	1	1	1	1	1
U20	1	1	1	1	1	1	1	1	1	1	1	1	1	1	1	1	1	1	1	1	1	1	1	1
U21	0	0	0	0	0	0	0	0	0	1	1	1	1	1	1	1	1	1	1	1	1/0	0	0	0
U22	0	0	0	0	0	0	0	0	0	0	0	0	0	0	0	0	0/1	1	1	1	1	1	1	0
U23	0	0	0	0	0	0	0	0	0	0	0	0	0	0	0	0	0	0/1	0/1	0/1	0/1	0/1	0	0
U24	1	1	1	1	1	1	1	1	1	1	1	1	1	1	1	1	1	1	1	1	1	1	1	1
U25	1	1	1	1	1	1	1	1	1	1	1	1	1	1	1	1	1	1	1	1	1	1	1	1
U26	1	1	1	1	1	1	1	1	1	1	1	1	1	1	1	1	1	1	1	1	1	1	1	1

This table presents the scheduling results of scenarios A and B. Where the results of two scenarios are the same, the values are shown in the corresponding cells. For those that are different, they are given in the same cell separated by "/", standing for "value of scenario A/value of scenario B".

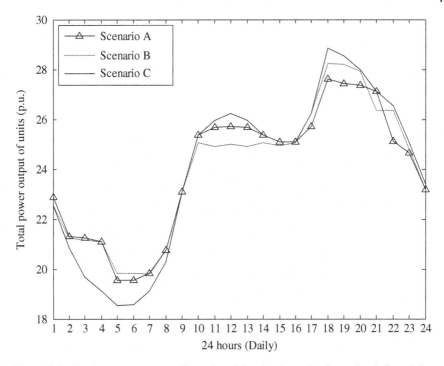

Figure 5.2 Total power output profiles of traditional units under Scenarios A, B, and C.

Eq. (5.10) is illustrated in Figure 5.3. It is clearly shown that BESS takes a substantial proportion of the system spinning reserve. Especially, BESSs relieve the duty of providing spinning reserve from the thermal units during peak-load periods.

The SCUC problem with BESSs for both peak-load shaving and reserve support is solved under different BESS penetration levels[1] to quantify the influences of different BESS capacities on the total production costs. The results are listed in Table 5.3. It can be seen that both SCs and OCs are reduced with the increasing capacity of BESS. Although the SCs are almost the same when the penetration level is at 8–10%, the OCs are still slightly decreasing.

Figure 5.4 depicts the result of sensitivity analysis of benefit with different levels of BESS penetration. The vertical axis represents the production cost reduction with a 1% BESS capacity increment and the horizontal axis is the BESS penetration

1 The BESS penetration level is the ratio of total BESS power to peak load of the power system.

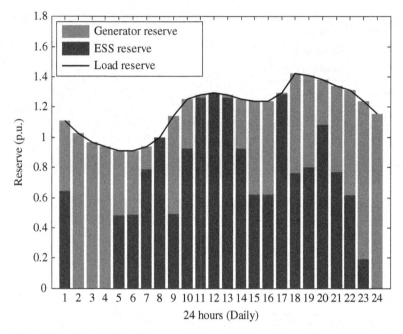

Figure 5.3 Composition of the operating reserve in each hour (Scenario A).

Table 5.3 Production costs under different BESS penetration levels.

BESS Penetration level	Startup cost	Operation cost	Total cost
0%	3240	758 684.73	761 924.73
1%	3210	755 417.73	758 627.73
2%	3150	753 414.30	756 564.30
3%	3190	752 353.48	755 543.48
4%	3170	751 067.00	754 237.00
5%	3100	749 938.93	753 038.93
6%	2940	749 001.72	751 941.72
7%	2900	747 859.60	750 759.60
8%	2850	747 063.21	749 913.21
9%	2850	746 234.09	749 084.09
10%	2850	745 638.07	748 488.07

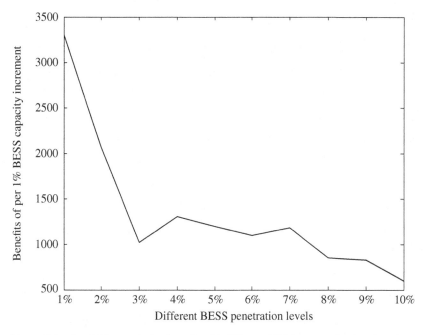

Figure 5.4 Sensitivity analysis of profit with different BESS penetration levels.

level. It can be seen that the benefit is very sensitive to the level of penetration and the marginal contribution of BESS decreases with the increase of BESS capacity.

5.3.3.4 Annual Benefits of Using BESSs

In order to calculate the annual benefit of using BESSs for the test system, the annual load profile of this system, taken from reference [16], is analyzed first. The daily load curves are clustered into 10 typical curves using a fast global k-means algorithm [15]. Then we solve the SCUC problem against the 10 typical daily loads under the three scenarios described above. The BESSs are the same as the base case, i.e. five BESSs, each with 40 MW rated power and 200 MWh rated energy capacity. Simulation results are shown in Table 5.4.

With 200 MW BESSs installed under Scenarios A and B, the annual economic benefit under Scenario B is $1 191 837.04, which is about 0.68% of the annual total OC without BESS, while the benefit under Scenario A is $1 857 399.66, which is 1.06% of the annual OC without BESS. This result reveals the merit of dispatching BESSs for both energy shifting and reserve support. However, the benefit of using BESSs is still relatively low compared to the currently high investment cost for deploying BESSs.

Table 5.4 Economic analyses under different typical load scenarios.

		Scenario A	Scenario B	Scenario C
Load scenarios	Duration time (day)	Total cost ($)	Total cost ($)	Total cost ($)
1	48	470 116.65	471 340.02	474 591.99
2	10	763 879.55	766 243.15	774 998.80
3	2	264 951.79	265 072.66	265 760.87
4	37	366 579.73	367 418.50	368 256.70
5	69	597 747.54	602 015.42	607 944.13
6	25	411 126.06	411 754.75	413 097.16
7	52	540 422.87	543 482.65	548 983.96
8	39	435 021.52	435 635.58	437 321.74
9	54	444 333.12	445 149.26	447 198.98
10	28	314 912.80	315 434.20	315 814.26

5.4 Stochastic and Robust SCUC with Energy Storage System and Wind Power

As mentioned in the introduction of this chapter, many research papers have been published on the stochastic SCUC problem considering the increasing penetration of wind and solar power generation. Energy storage system is typically a preferable resource to counteract the variation of REG power ouput. In this section, we will discuss two types of stochastic UC formulation with ESS.

5.4.1 Scenario-Based Stochastic SCUC

Each scenario is represented by s with an occurring possibility ρ_s, and ψ_{sc} stands for the scenario set. The objective function can be formulated as (5.19). It is composed of two parts. The first part is the unit startup cost and the second part is the expected production cost of all units considering all selected scenarios:

$$\min \sum_{t \in \psi_t, i \in \psi_g} [S_i u_{i,t}(1 - u_{i,t-1})] + \sum_{s \in \psi_{sc}} \rho_s \sum_{t \in \psi_t, i \in \psi_g} [u_{i,t} F_i(p_{i,s,t})] \qquad (5.19)$$

The BESS model given by Eqs. (5.1) to (5.7) should be modified as follows to consider all the scenarios:

$$E_{m,s,t} = E_{m,s,t-1} + \left(p^c_{em,s,t-1} \times \eta_{mc} - p^d_{em,s,t-1}/\eta_{md}\right) \times \Delta t, \forall t \in \psi_t, s \in \psi_{sc} \qquad (5.20)$$

$$0 \le p_{em,s,t}^c \le u_{em,s,t}^c \times p_{em}^{c\max}, \forall t \in \psi_t, s \in \psi_{sc} \tag{5.21}$$

$$0 \le p_{em,s,t}^d \le u_{em,s,t}^d \times p_{em}^{d\max}, \forall t \in \psi_t, s \in \psi_{sc} \tag{5.22}$$

$$E_m^{\min} \le E_{m,s,t} \le E_m^{\max}, \forall t \in \psi_t, s \in \psi_{sc} \tag{5.23}$$

$$0 \le u_{em,s,t}^c + u_{em,s,t}^d \le 1, \forall t \in \psi_t, s \in \psi_{sc} \tag{5.24}$$

$$0 \le p_{em,s,t}^r + p_{em,s,t}^d - p_{em,s,t}^c \le p_m^{d\max}, \forall t \in \psi_t, s \in \psi_{sc} \tag{5.25}$$

$$E_m^{\min} \le E_{m,s,t} - \left(p_{em,s,t}^r + p_{em,s,t}^d - p_{em,s,t}^c\right)/\eta_d \times \Delta t, \forall t \in \psi_t, s \in \psi_{sc} \tag{5.26}$$

1) Power balance constraints

The total load demand and power injection should be balanced for each scenario. In Eq.(5.27), the wind power $p_{wj,s,t}$ and the total load power $P_{d,s,t}$ under scenario s are both considered:

$$\sum_{i=1}^N p_{i,s,t} - \sum_{m=1}^M \left(p_{em,s,t}^c - p_{em,s,t}^d\right) + \sum_{j=1}^W p_{wj,s,t} = P_{d,s,t}, \forall t \in \psi_t, s \in \psi_{sc} \tag{5.27}$$

where W is the total number of wind farms.

2) System operating reserve requirement

The reserve requirement should be satisfied for each scenario considering the wind power variations:

$$\sum_{i=1}^N u_{i,t} P_{i,\max} - \sum_{m=1}^M \left(p_{em,s,t}^c - p_{em,s,t}^d - p_{em,s,t}^r\right) \ge (1+\rho)P_{d,s,t} + P_{w,s,t}^r, \forall t \in \psi_t, s \in \psi_{sc} \tag{5.28}$$

where $P_{w,s,t}^r$ is the preset reserve power at time t to deal with wind power fluctuations.

3) Generating unit constraints
- Power output limits

$$u_{i,t} \times P_i^{\min} \le p_{i,s,t} \le u_{i,t} \times P_i^{\max}, \forall i \in \psi_g, t \in \psi_t, s \in \psi_{sc} \tag{5.29}$$

- Ramping up/down limits

$$p_{i,s,t} - p_{i,s,t-1} \le u_{i,t-1} \times RU_i + P_i^{\max} \times (1 - u_{i,t-1}), \forall i \in \psi_g, t \in \psi_t, s \in \psi_{sc} \tag{5.30}$$

$$p_{i,s,t} - p_{i,s,t-1} \ge -u_{i,t-1} \times RD_i - P_i^{\max} \times (1 - u_{i,t}), \forall i \in \psi_g, t \in \psi_t, s \in \psi_{sc} \tag{5.31}$$

- Minimum up/down time limits
  The constraints considering the unit minimum up and down times are the same as the deterministic UC formulation given in Section 5.3:

$$\sum_{k=t-SU_i+1}^{t} su_{i,k} \leq u_{i,t}, \forall i \in \psi_g, t \in \psi_t \tag{5.32}$$

$$\sum_{k=t-SD_i+1}^{t} sd_{i,k} \leq 1 - u_{i,t}, \forall i \in \psi_g, t \in \psi_t \tag{5.33}$$

4) State switching limits of each BESS
   The following constraints mean that the total times of state switching are counted and capped for each scenario. For other types of ESS, e.g. compressed air energy storage, the charging and discharging status for each scenario may be set the same for all the scenarios in the UC formulation.

$$\sum_{t=1}^{T} \left(\left| u^c_{em,s,t} - u^c_{em,s,t-1} \right| + \left| u^d_{em,s,t} - u^d_{em,s,t-1} \right| \right) \leq k^{\max}_m, \forall m \in \psi_{es}, s \in \psi_{sc} \tag{5.34}$$

5) Constraints on residue energy of each BESS

   The following constraints limit the allowable residue energy difference within a dispatch cycle:

$$\left| E_{m,0} - E_{m,s,T} \right| \leq \varepsilon E_{m,0}, \forall m \in \psi_{es}, s \in \psi_{sc} \tag{5.35}$$

6) Other constraints on each BESS

   For each BESS, equations (5.20)–(5.26) should be included.

7) DC power flow limits

$$-F^{\max}_l \leq \sum_{i \in \psi_n} K_{l,i} \left(\sum_{j \in \psi_{gi}} P_{j,s,t} - P_{di,s,t} + p^d_{ei,s,t} - p^c_{ei,s,t} + P_{wi,s,t} \right) \leq F^{\max}_l, \forall l \in \psi_l, s \in \psi_{sc} \tag{5.36}$$

For the scenario-based stochastic SCUC formulation, the problem type is the same as that of the deterministic SCUC. Therefore, it can be solved by the mature optimization algorithm or available optimization software.

5.4.2 Robust SCUC

Different from the stochastic optimization approach, the uncertain parameters in a robust optimization formulation do not follow probabilistic distributions, but rather belong to intervals or sets. For a robust SCUC formulation, intervals are generally used to represent the uncertain power injections/absorptions, e.g. wind power and load demands. If only the wind power prediction error is considered, the uncertainty set can be described as [9]

$$\Lambda := \left\{ \mathbf{p}_w \in R^{W \times T} : \sum_{t=1}^{T} \left(z_{wj,t}^{+} + z_{wj,t}^{-} \right) \leq \Gamma_{wj}, P_{wj,t} = \overline{P}_{wj,t} + z_{wj,t}^{+} p_{wj,t}^{u} - z_{wj,t}^{-} p_{wj,t}^{d}, \forall t \in \psi_t, \forall j \in \psi_w \right\}$$

(5.37)

where $\overline{p}_{wj,t}$ is the forecasted wind power output, the interval $[-p_{wj,t}^{d}, p_{wj,t}^{u}]$ means the range of wind power forecast error, which can be obtained by historical data, $z_{wj,t}^{+}$ and $z_{wj,t}^{-}$ are binary variables to model the wind power deviations, and Γ_{wj} is called the "budget of uncertainty." When $\Gamma_{wj} = 0$, it is equivalent to the deterministic case. As Γ_{wj} increases, the size of Λ extends.

With this uncertainty set definition, we can formulate the following objective function of the robust SCUC problem [9, 17]:

$$\min_{\mathbf{p},\mathbf{u}} \left\{ \sum_{t \in \psi_t} \sum_{i \in \psi_g} [S_i u_{i,t}(1 - u_{i,t-1})] + \max_{\mathbf{p}_w \in \Lambda} \sum_{t \in \psi_t} \sum_{i \in \psi_g} [u_{i,t} F_i(p_{i,t})] \right\}$$

(5.38)

where the vectors $\mathbf{p}$ and $\mathbf{u}$ include all the power output and state variables of generating units and BESSs. The constraints (5.1) to (5.7) and (5.11) to (5.17) are all the same and should be included. The constraints (5.9), (5.10), and (5.18) should be rewritten as follows:

$$\sum_{i=1}^{N} p_{i,t} - \sum_{m=1}^{M} \left(p_{em,t}^{c} - p_{em,t}^{d} \right) + \sum_{j=1}^{W} P_{wj,t} = P_{d,t}, \forall t \in \psi_t$$

(5.39)

$$\sum_{i=1}^{N} u_{i,t} P_i^{\max} - \sum_{m=1}^{M} \left(p_{em,t}^{c} - p_{em,t}^{d} - p_{em,t}^{r} \right) \geq (1 + \gamma) P_{d,t} + P_{w,t}^{r}, \forall t \in \psi_t$$

(5.40)

$$-F_l^{\max} \leq \sum_{i \in \psi_n} K_{l,i} \left(\sum_{j \in \psi_{gi}} p_{j,t} - P_{di,t} + p_{ei,t}^{d} - p_{ei,t}^{c} + P_{wi,t} \right) \leq F_l^{\max}, \forall l \in \psi_l$$

(5.41)

It should be noted that $p_{wj,t}$ is a variable here. In order to solve the robust SCUC problem, the objective function can be transformed into the following form:

$$\min_{\mathbf{u}} \left\{ \sum_{t \in \psi_t} \sum_{i \in \psi_g} [S_i u_{i,t}(1 - u_{i,t-1})] + \max_{\mathbf{p}_w \in \Lambda} \min_{\mathbf{p} \in \Omega(\mathbf{u}, \mathbf{p}_w)} \sum_{t \in \psi_t} \sum_{i \in \psi_g} [u_{i,t} F_i(p_{i,t})] \right\}$$

(5.42)

where $\mathbf{p} \in \Omega(\mathbf{u}, \mathbf{p}_w)$ is the set of feasible power dispatch solutions for units with a fixed commitment decision $\mathbf{u}$ and wind power $\mathbf{p}_w$. For this optimization formulation, the related constraints should all be considered. The inner or the second stage problem is now in a max–min form. Using dual theory, the max–min form can be changed to a max–max problem, which is actually a maximization problem and can be readily solved. The two-stage optimization problem can be iteratively solved by Benders decomposition method. Refer to references [9] and [17] for detailed derivations.

5.5 Conclusion and Discussion

In this chapter, we first discussed the model of BESS for the UC problem. Then a deterministic SCUC formulation with thermal units and BESSs is illustrated. Simulations on the modified IEEE RTS 24 bus system show the benefits of BESSs for SCUC. The results indicate that a lower UC cost can be achieved by dispatching BESSs for both peak-load shaving and reserve support. The formulations of scenario-based stochastic and robust SCUCs are derived by considering uncertain power injections. It should be noted that the inclusion of ESS will not bring difficulties or challenges for the SCUC problems. The most important point is to build an appropriate model for the ES evolved.

References

1 Wood, A.J. and Wollenberg, B.F. (2012). *Power Generation, Operation, and Control*. Wiley.

2 Sheble, G.B. and Fahd, G.N. (1994). Unit commitment literature synopsis. *IEEE Transactions on Power Systems* 9 (1): 128–135.

3 Padhy, N.P. (2004). Unit commitment – A bibliographical survey. *IEEE Transactions on Power Systems* 19 (2): 1196–1205.

4 Zheng, Q.P., Wang, J., and Liu, A.L. (2015). Stochastic optimization for unit commitment – A review. *IEEE Transactions on Power Systems* 30 (4): 1913–1924.

5 Li, T. and Shahidehpour, M. (2005). Price-based unit commitment: A case of Lagrangian relaxation versus mixed integer programming. *IEEE Transactions on Power Systems* 20 (4): 2015–2025.

6 Hobbs, B.F., Rothkopf, M.H., O'Neill, R.P., and Chao, H.P. (eds.) (2001). *The Next Generation of Electric Power Unit Commitment Models*, vol. 36. Springer Science & Business Media.

7 Senjyu, T., Miyagi, T., Yousuf, S.A. et al. (2007). A technique for unit commitment with energy storage system. *International Journal of Electrical Power and Energy Systems* 29 (1): 91–98.

8 Daneshi, H. and Srivastava, A.K. (2012). Security-constrained unit commitment with wind generation and compressed air energy storage. *IET Generation, Transmission and Distribution* 6 (2): 167–175.

9 Jiang, R., Wang, J., and Guan, Y. (2012). Robust unit commitment with wind power and pumped storage hydro. *IEEE Transactions on Power Systems* 27 (2): 800–810.

10 Pozo, D., Contreras, J., and Sauma, E.E. (2014). Unit commitment with ideal and generic energy storage units. *IEEE Transactions on Power Systems* 29 (6): 2974–2984.

11 Wen, Y., Guo, C., Pandžić, H. et al. (2016). Enhanced security-constrained unit commitment with emerging utility-scale energy storage. *IEEE Transactions on Power Systems* 31 (1): 652–662.

12 Bruninx, K., Dvorkin, Y., Delarue, E. et al. (2016). Coupling pumped hydro energy storage with unit commitment. *IEEE Transactions on Sustainable Energy* 7 (2): 786–796.

13 Guena, T. and Leblanc, P. (2006). How depth of discharge affects the cycle life of lithium-metal-polymer batteries. *Proceedings of the Telecommunications Energy Conference*, pp. 1–8.

14 Carrión, M. and Arroyo, J.M. (2006). A computationally efficient mixed-integer linear formulation for the thermal unit commitment problem. *IEEE Transactions on Power Systems* 21 (3): 1371–1378.

15 Likas, A., Vlassis, N., and Verbeek, J.J. (Februsry 2003). The global k-means clustering algorithm. *Pattern Recognition* 36 (2): 451–461.

16 Reliability Test System Task Force (August 1999). The IEEE reliability test system-1996. *IEEE Transactions on Power Systems* 14 (3): 1010–1020.

17 Bertsimas, D., Litvinov, E., Sun, X.A. et al. (2013). Adaptive robust optimization for the security constrained unit commitment problem. *IEEE Transactions on Power Systems* 28 (1): 52–63.

6

Optimal Power Flow with Energy Storage System

6.1 Introduction

Optimal power flow (OPF) is one of the most widely studied nonlinear optimization problems in the power system community since it was introduced by Carpentier in 1962 [1]. It aims at finding a least-cost schedule with the operational constraints satisfied, which acts as an important tool for economical and secure operation of power systems. Mathematically, OPF is a large-scale nonlinear and nonconvex optimization problem, which is non-deterministic polynomial (NP) hard and difficult to find the global optimal solution. It should be pointed out that the OPF problem we will discuss in this chapter is the so-called alternating current (AC) OPF problem for the meshed transmission networks.

Many methods have been proposed to solve OPF problems, mainly including nonlinear programming methods and modern heuristic (artificial intelligent) approaches. These methods can be classified into deterministic, non-deterministic and hybrid methods in references [1] and [2]. The non-deterministic methods discussed in reference [2] are actually the various modern heuristic methods, e.g., genetic algorithm, particle swarm optimization, ant colony optimization, and artificial neural networks. Theoretically, these heuristic methods have the capabilities of finding global optimal solutions. However, they cannot guarantee to pinpoint the global optimum and often require extensive computation, especially for large-scale power systems. Regarding the nonlinear programming methods, the interior point method (IPM) is the most famous and extensively studied method [3]. Although the IPM was first proposed to solve the linear programming problem, it has been successfully applied to solve the nonlinear OPF problems.

When an OPF solution is found by an IPM, we do not know whether it is the global optimum. Therefore, with the advancement of nonlinear convex optimization, it has been a hot research area to use convex relaxation techniques, especially semidefinite programming (SDP) relaxation, to solve OPF problems in recent years [4, 5]. The conditions on the exactness of convex relaxation have been proved and

Energy Storage for Power System Planning and Operation, First Edition. Zechun Hu.
© 2020 John Wiley & Sons Singapore Pte. Ltd.
Published 2020 by John Wiley & Sons Singapore Pte. Ltd.

the effectiveness of these methods to get global optimum for the test and practical systems have been extensively tested. Normally, the computational burden to solve the OPF problem by SDP-based methods is much higher than that of using IPMs directly on the non-convexified OPF formulation.

For a transmission system installed with one or multiple large-scale energy storage systems, the OPF problem should take the energy constraints of each ESS into consideration. The straightforward approach is to extend the OPF problem into a multi-period optimization problem [6]. However, the size of the OPF problem is increased, which is proportional to the number of time periods.

In this chapter, we will first build a mathematical formulation of a basic OPF problem with ESS. Then solving the problem directly using IPM and transforming the problem into an SDP will be introduced, respectively. Finally, simulation results using the two types of methods will be illustrated and discussed.

6.2 Optimal Power Flow Formulation with Energy Storage System

6.2.1 Multi-Period OPF and Rolling Optimization

Classical OPF solves the optimal dispatch problem with a given load demand of each bus at a specific time. However, as the energy balance constraints for an ESS are temporarily coupled, the OPF for a power system with at least one ESS can be naturally extended from a single-period to a multi-period optimization problem. For practical applications, the multi-period OPF problem can be solved periodically with a preset time interval using updated load forecast results and other updated information. Figure 6.1 illustrates the rolling optimization flow-chart of the multi-period OPF problem. The time span of the OPF problem can be 24 hours or even longer. Starting from $t = T_0$, the OPF problem will be run to obtain the optimal dispatch results for the time span of T with a time interval Δt. The generators' outputs, voltage setting points, and the charging/discharging schedule of each ESS will be dispatched by the system operator to execute for the time interval $[t, t + \Delta t]$. At the time $t + \Delta t$, the multi-period OPF problem within $[t + \Delta t, T_f]$ will be solved again using updated data. When the time approaches $t = T_f$, another new optimization cycle will be started.

6.2.2 Energy Storage Model for the OPF Problem

In this chapter, we do not consider the discrete decision variables, e.g. the position of the tap changer and on/off status of the shunt capacitor, in the OPF problem for simplicity. Therefore, the binary variables denoting the charging and discharging states of the ESS are also ignored. We assume that this will not affect the

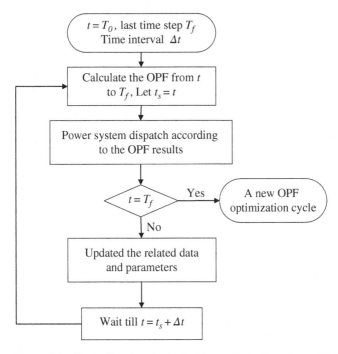

Figure 6.1 Illustrative flowchart of rolling optimization of a multi-period OPF problem with ESS.

optimization results for the ESS according to the research results of references [7] and [8]. Thus, different from the ESS model for SCUC given in Chapter 5, the following model will be used for the OPF problem:

$$E_{m,t} = E_{m,t-1} + \left(P^c_{em,t-1} \times \eta_{mc} - P^d_{em,t-1}/\eta_{md}\right) \times \Delta t, \ \forall m \in \psi_{es}, t \in \psi_t \tag{6.1}$$

$$P^{c\,\min}_{em} \leq P^c_{em,t} \leq P^{c\,\max}_{em}, \ \forall m \in \psi_{es}, t \in \psi_t \tag{6.2}$$

$$P^{d\,\min}_{em} \leq P^d_{em,t} \leq P^{d\,\max}_{em}, \ \forall m \in \psi_{es}, t \in \psi_t \tag{6.3}$$

$$E^{\min}_m \leq E_{m,t} \leq E^{\max}_m, \ \forall m \in \psi_{es}, t \in \psi_t \tag{6.4}$$

Furthermore, the reactive power support from ESS is also taken into account. The reactive power constraints are as follows:[1]

$$Q^{\min}_{em} \leq Q_{em,t} \leq Q^{\max}_{em}, \ \forall m \in \psi_{es}, t \in \psi_t \tag{6.5}$$

---

1 Here, the constraints on the range reactive power of ESS is very simple. Other types of constraints can also be formulated, e.g., by limiting the current or apparent power of ESS.

Considering a dispatch cycle, e.g. 24 hours, with given initial residue energy $E_{m,0}^{sp}$ for ESS, the residue energy at the end of the cycle should be within an interval as shown in (6.7):

$$E_{m,0} = E_{m,0}^{sp} \tag{6.6}$$

$$E_{m,0}^{sp} - \varepsilon_m \le E_{m,T} \le E_{m,0}^{sp} + \varepsilon_m \tag{6.7}$$

$$0 \le \varepsilon_m \le \varepsilon_m^{sp} \tag{6.8}$$

Here, ε_m is a slack variable for the energy balance of ESS and its maximum value is the specified value ε_m^{sp}.

6.2.3 OPF Formulation

We consider an N-bus power system with G generator buses ($G \le N$) and M buses connected with ESSs ($M \le N$). For simplicity, we assume each generator bus only has one generator or an aggregated generator and each bus with ESS only has one energy storage system. Define the set of all buses $\psi_n := \{1, ..., N\}$, the set of generator buses $\psi_g := \{1, ..., G\}$, and the set of buses with ESSs $\psi_{es} := \{1, ..., M\}$. The admittance matrix of this power system is represented by $Y \in C^{n \times n}$.

For each generator bus $k \in \psi_g$, the active and reactive power outputs should be within the allowable range as follows:

$$P_{gk}^{min} \le P_{gk,t} \le P_{gk}^{max}, \forall k \in \psi_g, t \in \psi_t \tag{6.9}$$

$$Q_{gk}^{min} \le Q_{gk,t} \le Q_{gk}^{max}, \forall k \in \psi_g, t \in \psi_t \tag{6.10}$$

For each bus $k \in \psi_n$, the magnitude of nodal voltage $\dot{V}_k(t)$ should satisfy the following constraint:

$$V_k^{min} \le |\dot{V}_{k,t}| \le V_k^{max}, \forall t \in \psi_t \tag{6.11}$$

For the nodal power balance equations, they are usually formed as follows in polar coordinates:

$$P_{gk,t} - P_{dk,t} - P_{ek,t}^c + P_{ek,t}^d = V_{k,t} \sum_{j:(kj) \in \psi_b} V_{j,t} \left[G_{kj} \cos\left(\delta_{kj,t}\right) + B_{ij} \sin\left(\delta_{kj,t}\right) \right], \forall k \in \psi_n, t \in \psi_t \tag{6.12}$$

$$Q_{gk,t} - Q_{gk,t} - Q_{dk,t} - Q_{ek,t} = V_{k,t} \sum_{j:(kj) \in \psi_b} V_{j,t} \left[G_{kj} \sin\left(\delta_{kj,t}\right) - B_{kj} \cos\left(\delta_{kj,t}\right) \right], \forall k \in \psi_n, t \in \psi_t \tag{6.13}$$

Here, we use the convention that for a non-generator bus $k \in \psi_n \backslash \psi_g$, $P_{gk,t} = Q_{gk,t} = 0$. For a bus connected with ESS $k \in \psi_{es}$, the constraints (6.1) to (6.8) should be satisfied, and for a bus without ESS, $k \in \psi_n \backslash \psi_{es}$, $P_{ek,t} = Q_{ek,t} = 0$.

To limit the power flow from a branch, the constraints can be in the form of a branch current, active power flow, or apparent power. The constraints of the apparent power of the branch from bus k to j are shown as follows:

$$\left| S_{kj,t}(\boldsymbol{V}_t, \boldsymbol{\theta}_t) \right| \le S_{kj}^{\max}, \ \forall(kj) \in \psi_b, t \in \psi_t \tag{6.14}$$

where $\boldsymbol{V}_t$ is the vector of nodal voltage magnitude and $\boldsymbol{\theta}_t$ is the vector of a nodal voltage phase angle.

Based on the above analysis, the OPF formulation with ESSs can be obtained:

$$\varphi := \min_{\boldsymbol{V}_t, \boldsymbol{\theta}_t, \boldsymbol{P}_{g,t}, \boldsymbol{Q}_{g,t}, \boldsymbol{P}_{e,t}^c, \boldsymbol{P}_{e,t}^d, \boldsymbol{Q}_{e,t}} \left[\sum_{t \in \psi_t} \sum_{k \in \psi_g} f_k(P_{k,t}) + \sum_{k \in \psi_{es}} s_k \varepsilon_k \right] \tag{6.15}$$

s.t. Eqs. (6.1) to (6.14)

where $f_k(\cdot)$ is the production cost function of generator bus k and the quadratic function

$$f_k(P_{k,t}) = c_{k2}(P_{k,t})^2 + c_{k1}P_{k,t} + c_{k0}$$

is commonly used. The second item in the OPF objective function $s_k \varepsilon_k$ represents the penalty to the residue energy deviation of ESS k during the optimization cycle. As shown in Figure 6.1, the penalty efficiency s_k can be adjusted and updated after each optimization according to the optimization results. For example, s_k can be calculated based on the average generation cost reduction of each kilowatt-hour of energy absorption or release from an ESS.

It should be emphasized that the above OPF formulation is a very basic one. Other objective functions and constraints can be built and added to the formulation.

6.3 Interior Point Method to Solve the Multi-Period OPF Problem

6.3.1 Optimal Condition for the Interior Point Method

The above OPF formulation can be written in the form of a general nonlinear programming problem as follows:

$$\begin{cases} \min \ f(\boldsymbol{x}) \\ \text{s.t.} \\ \quad g(\boldsymbol{x}) = 0 \\ \quad h(\boldsymbol{x}) \le 0 \end{cases} \tag{6.16}$$

where $\boldsymbol{x}$ is a vector that includes both control variables and state variables (voltage magnitude and phase angle of all buses). In order to solve the problem using the interior point method, the inequality constraints are transformed into equality constraints by introducing non-negative slack variables:

$$h(x) + \zeta = 0 \tag{6.17}$$

$$\zeta \geq 0 \tag{6.18}$$

The inequality constraints are further eliminated by adding logarithmic barrier terms to the objective function. Then the following optimization problem with only equality constraints can be obtained:

$$\begin{cases} \min \ f(x) - \mu \sum_{i=1}^{N_{ieq}} \ln \zeta_i \\ \text{s.t.} \\ \quad g(x) = 0 \\ \quad h(x) + \zeta = 0 \end{cases} \tag{6.19}$$

where N_{ieq} is the number of inequality constraints in Eq. (6.16) and μ is a positive scalar called the barrier parameter.

The Lagrangian function can be formulated as follows, which transforms the equality constrained optimization problem into an unconstrained one:

$$L(y) = f(x) - \mu \sum_{i=1}^{N_{ieq}} \ln \zeta_i - \lambda^T g(x) - \pi^T [h(x) + \zeta] \tag{6.20}$$

where $y = [x, \zeta, \lambda, \pi]^T$ and λ and π are the vectors of Lagrange multipliers, which are called dual variables.

Now, the first-order necessary optimality conditions, or the Karush–Kuhn–Tucker (KKT) conditions of the above problem can be derived by setting the derivatives of Eq. (6.20) with respect to y to zero [3]:

$$L_x = \frac{\partial L(y)}{\partial x} \equiv \nabla_x f(x) - \nabla_x g(x)^T \lambda - \nabla_x h(x)^T \pi = 0 \tag{6.21}$$

$$L_\zeta = \frac{\partial L(y)}{\partial \zeta} = \pi - \mu \cdot diag\{\zeta^{-1}\}e \Rightarrow L_\zeta^\mu = diag\{\zeta\}\pi - \mu e = 0 \tag{6.22}$$

$$L_\lambda = \frac{\partial L(y)}{\partial \lambda} \equiv g(x) = 0 \tag{6.23}$$

$$L_\pi = \frac{\partial L(y)}{\partial \pi} \equiv h(x) + \zeta = 0 \tag{6.24}$$

where $e = [1, \ldots, 1]^T$. From (6.22), we can calculate the "average" μ with given ζ and π as follows:

$$\mu = \frac{\zeta^T \pi}{N_{ieq}} \tag{6.25}$$

Define $\rho \equiv \zeta^T \pi$, which is called the complementarity gap. It is proved that when $\rho \to 0$, $\mu \to 0$, the solution $x(\mu)$ converges to a local optimum of the problem (6.19).

6.3.2 Procedure of the Primal-Dual IPM to Solve the OPF Problem

The procedure of the primal-dual IPM to solve the OPF problem can be found in references [3] and [9]. Here, we briefly outline the steps of the primal-dual IPM to solve Eqs. (6.21) to (6.24).

Step 1: Initialization. Set the initial values $y^{(0)}$ and $\mu^{(0)}$ and the iteration counter $\kappa = 0$.

Step 2: Solve the linearized KKT conditions for $\Delta y^{(\kappa)}$. Form the Jacobian matrix $J_{KKT}(y^{(\kappa)})$ of Eqs. (6.21) to (6.24) and solve the following equations:

$$
J_{KKT}\left(y^{(\kappa)}\right)
\begin{bmatrix}
\Delta x^{(\kappa)} \\
\Delta \zeta^{(\kappa)} \\
\Delta \lambda^{(\kappa)} \\
\Delta \pi^{(\kappa)}
\end{bmatrix}
=
\begin{bmatrix}
-\nabla_x f\left(x^{(\kappa)}\right) + \nabla_x g\left(x^{(\kappa)}\right)^T \lambda + \nabla_x h\left(x^{(\kappa)}\right)^T \pi \\
diag\left\{\zeta^{(\kappa)}\right\}\pi^{(\kappa)} - \mu^{(\kappa)} e \\
-g\left(x^{(\kappa)}\right) \\
-h\left(x^{(\kappa)}\right) - \zeta^{(\kappa)}
\end{bmatrix}
$$

$$(6.26)$$

where

$$
J_{KKT}\left(y^{(\kappa)}\right) =
\begin{bmatrix}
H_{KKT}^{(\kappa)} & 0 & -\nabla_x g\left(x^{(\kappa)}\right) & -\nabla_x h\left(x^{(\kappa)}\right) \\
0 & diag\left\{\pi^{(\kappa)}\right\} & 0 & diag\left\{\zeta^{(\kappa)}\right\} \\
\nabla_x g\left(x^{(\kappa)}\right) & 0 & 0 & 0 \\
\nabla_x h\left(x^{(\kappa)}\right) & I & 0 & 0
\end{bmatrix}
$$

$$(6.27)$$

and $H_{KKT}^{(\kappa)} = \nabla_x^2 f\left(x^{(\kappa)}\right) - \nabla_x^2 g\left(x^{(\kappa)}\right)^T \lambda - \nabla_x^2 h\left(x^{(\kappa)}\right)^T \pi$.

Step 3: Update variables. The variables are updated by

$$
y^{(k+1)} = y^{(k)} + \alpha^{(k)} \Delta y^{(k)}
$$

$$(6.28)$$

The method to choose a proper $\alpha^{(k)}$ is discussed in references [3] and [9].

Step 4: Check the convergence conditions. In reference [9], the iteration terminates when the complementarity gap falls below the preset threshold, while in reference [3], the primal feasibility, dual feasibility, and the objective function variation between the two successive iterations are also checked to ensure that an optimal solution is found. If convergence has not been achieved, the barrier parameter is updated by Eq. (6.29); set $\kappa = \kappa + 1$ and go to Step 2.

$$\mu^{(\kappa)} = \sigma \frac{\rho^{(\kappa)}}{N_{ieq}} \tag{6.29}$$

where $\sigma \in (0, 1)$ is called the center parameter and is typically set between 0.1 and 0.2.

6.3.3 Discussion on Singularities Caused by Constraints of Energy Storage System

The IPM is successfully used to solve the OPF problems under a fixed load demand for one time step. For the formulation given in Section 6.2.3, it is an OPF problem with multiple time steps. The time-dependent constraint corresponding to ESS may cause the matrix (6.27) to become singular as the IPM iterates toward an optimal solution. This problem is discussed in detail in references [10] and [11]. Different approaches are proposed in reference [10] to avoid the singularity problem using the Moore-Penrose pseudoinverse method, considering ESS standby losses, or removing the binding intertemporal constraints. However, the computational costs and reliability of each approach are different. The approach to consider ESS standby losses is adopted in reference [11] because it can be implemented easily.

6.4 Semidefinite Programming for the OPF Problem

Because constraints (6.11), (6.12) to (6.14) are non-convex, the OPF problem (6.15) is a non-convex optimization problem. Although the IPMs have been widely used to solve the OPF problems, the global optimality of the solution obtained by the IPMs cannot be guaranteed. With the development of convex optimization, using SDP methods to solve the OPF problems were proposed [12]. In order to derive the SDP formulation of an OPF problem, the nodal power balance equations are represented in the following form:

$$\dot{V}_{k,t}\dot{I}^*_{k,t} = P_{gk,t} - P_{dk,t} - P^c_{ek,t} + P^d_{ek,t} + \left[Q_{gk,t} - Q_{dk,t} - Q_{ek,t}\right]i, \quad \forall t \in \psi_t \tag{6.30}$$

where $\dot{I}_t = Y\dot{V}_t$ and i is the imaginary unit. For the buses without a generator $k \in \psi_n \backslash \psi_g$, $P_{gk,t} = Q_{gk,t} = 0$, while for the buses without ESS $k \in \psi_n \backslash \psi_{es}$, $P_{ek,t} = Q_{ek,t} = 0$.

In the following, we first give a convex relaxation of the optimization problem (6.15) according to the derivation of references [13] and [14]. Then the Lagrangian dual of the relaxation problem will be constructed and solved. It can be proved that the gap between the Lagrangian dual problem and Eq. (6.15) is zero under some specific conditions.

6.4.1 Convex Relaxation of the OPF Problem

We define the following variables, which are similar to those in references [13] and [14]:

$$M_k := \begin{bmatrix} e_k e_k^T & 0 \\ 0 & e_k e_k^T \end{bmatrix}$$

$$Y_k := e_k e_k^T Y$$

$$\mathbf{Y}_k := \frac{1}{2} \begin{bmatrix} \mathrm{Re}\{Y_k + Y_k^T\} & \mathrm{Im}\{Y_k^T - Y_k\} \\ \mathrm{Im}\{Y_k - Y_k^T\} & \mathrm{Re}\{Y_k + Y_k^T\} \end{bmatrix}$$

$$\overline{\mathbf{Y}}_k := -\frac{1}{2} \begin{bmatrix} \mathrm{Im}\{Y_k + Y_k^T\} & \mathrm{Re}\{Y_k - Y_k^T\} \\ \mathrm{Re}\{Y_k^T - Y_k\} & \mathrm{Im}\{Y_k + Y_k^T\} \end{bmatrix}$$

$$U_t := \begin{bmatrix} \mathrm{Re}\left(\dot{V}_t\right) \\ \mathrm{Im}\left(\dot{V}_t\right) \end{bmatrix}$$

$$W_t := U_t U_t^T$$

$$a_{k0,t} := c_{k1}\left[tr\{Y_k W_t\} + P_{ek,t}^c - P_{ek,t}^d + P_{dk,t}\right] + c_{k0} - \alpha_{k,t}, \forall k \in \psi_g$$

$$a_{k1,t} := \sqrt{c_{k2}}\left[tr\{Y_k W_t\} + P_{ek,t}^c - P_{ek,t}^d + P_{dk,t}\right], \forall k \in \psi_g$$

where $e_k (k = 1, ..., n)$ are the standard basis vectors in R^n. According to the derivation given in reference [13], the OPF problem (6.15) is equivalent to

$$\varphi = \min_{W_t, \alpha_t, E_t, P_{e,t}^c, P_{e,t}^d, Q_{e,t}} \left(\sum_{t \in \psi_t} \sum_{k \in \psi_g} \alpha_{k,t} + \sum_{k \in \psi_{es}} c_k \varepsilon_k \right) \tag{6.31}$$

s.t.

$$P_{gk}^{\min} - P_{dk,t} \leq tr\{Y_k W_t\} + P_{ek,t}^c - P_{ek,t}^d \leq P_{gk}^{\max} - P_{dk,t}, \forall k \in \psi_n, t \in \psi_t \tag{6.32}$$

$$Q_{gk}^{\min} - Q_{dk,t} \leq tr\{\overline{Y}_k W_t\} + Q_{ek,t} \leq Q_{gk}^{\max} - Q_{dk,t}, \forall k \in \psi_n, t \in \psi_t \tag{6.33}$$

$$\left(V_k^{\min}\right)^2 \leq tr\{M_k W_t\} \leq \left(V_k^{\max}\right)^2, \forall k \in \psi_n, t \in \psi_t \tag{6.34}$$

$$E_m^{\min} \leq E_{m,t} \leq E_m^{\max}, \forall m \in \psi_{es}, t \in \psi_t \tag{6.35}$$

$$P_{em}^{c\,\min} \leq P_{em,t}^c \leq P_{em}^{c\,\max}, \forall m \in \psi_{es}, t \in \psi_t \tag{6.36}$$

$$P_{em}^{d\,min} \leq P_{em,t}^{d} \leq P_{em}^{d\,max}, \quad \forall m \in \psi_{es}, t \in \psi_t \tag{6.37}$$

$$Q_{em}^{min} \leq Q_{em,t} \leq Q_{em}^{max}, \quad \forall m \in \psi_{es}, t \in \psi_t \tag{6.38}$$

$$E_{m,t} = E_{m,t-1} + \left(P_{em,t-1}^{c} \times \eta_{mc} - P_{em,t-1}^{d}/\eta_{md}\right) \times \Delta t, \quad \forall m \in \psi_{es}, t \in \psi_t \tag{6.39}$$

$$E_{m,0} = E_{m,0}^{sp}, \quad \forall m \in \psi_{es} \tag{6.40}$$

$$E_{m,0}^{sp} - \varepsilon_m \leq E_{m,T} \leq E_{m,0}^{sp} + \varepsilon_m, \quad \forall m \in \psi_{es} \tag{6.41}$$

$$0 \leq \varepsilon_m \leq \varepsilon_m^{sp}, \quad \forall m \in \psi_{es} \tag{6.42}$$

$$\begin{bmatrix} a_{k0,t} & a_{k1,t} \\ a_{k1,t} & -1 \end{bmatrix} \leq 0, \forall k \in \psi_g, t \in \psi_t \tag{6.43}$$

$$\boldsymbol{W}_t \geq 0, t \in \psi_t \tag{6.44}$$

$$rank(\boldsymbol{W}_t) = 1, \quad t \in \psi_t \tag{6.45}$$

It should be noted that constraint (6.43) is the equivalent form of $f_k(P_{k,t}) \leq \alpha_{k,t}$. The definition $\boldsymbol{W}_t := \boldsymbol{U}_t \boldsymbol{U}_t^T$ is replaced by constraints (6.44) and (6.45). The optimization problem (6.31) to (6.45) has a linear objective function and convex constraints except for constraint (6.45). If the rank one constraint is removed, then the problem (6.31) to (6.44) is a semidefinite program, which is a convex problem [13].

6.4.2 Lagrange Relaxation and Dual Problem

The Lagrange dual for the above SDP problem can be formed as follows:

$$\varphi^* = \max_{x \geq 0, z, \sigma, \beta, \Delta \geq 0} h(x, z, \sigma, \beta, \Delta) \tag{6.46}$$

s.t.

$$A_t \geq 0 \tag{6.47}$$

$$H_{k,t} + \xi_{k,t}^{max} - \xi_{k,t}^{min} = 0 \tag{6.48}$$

$$\Lambda_{k,t}(t) + \rho_{k,t}^{max} - \rho_{k,t}^{min} + \sigma_{k,t+1}\eta_{mc}\Delta t = 0 \tag{6.49}$$

$$-\Lambda_{k,t} + \theta_{k,t}^{max} - \theta_{k,t}^{min} + \sigma_{k,t+1}\Delta t/\eta_{md} = 0 \tag{6.50}$$

$$\gamma_{k,t}^{max} - \gamma_{k,t}^{min} + \sigma_{k,t+1} - \sigma_{k,t} = 0 \tag{6.51}$$

$$\gamma_{k,1}^{max} - \gamma_{k,1}^{min} + \sigma_{k,2} + \beta_k = 0 \tag{6.52}$$

$$\gamma_{k,T+1}^{max} - \gamma_{k,T+1}^{min} - \sigma_{k,T+1} - \delta_k^{min} + \delta_k^{max} = 0 \tag{6.53}$$

$$c_k - \delta_k^{min} - \delta_k^{max} - \delta_k = 0 \tag{6.54}$$

$$\begin{bmatrix} 1 & z_{l1,t} \\ z_{l1,t} & z_{l2,t} \end{bmatrix} \geq 0 \tag{6.55}$$

Here,

$$\Lambda_{k,t} := \begin{cases} \lambda_{k,t}^{\max} - \lambda_{k,t}^{\min} + c_{k1} + 2\sqrt{c_{k2}}z_{k1,t}, & \forall k \in \psi_g \\ \lambda_{k,t}^{\max} - \lambda_{k,t}^{\min}, & k \in \psi_n \backslash \psi_g \end{cases}$$

$$H_{k,t} := \zeta_{k,t}^{\max} - \zeta_{k,t}^{\min}, k \in \psi_n$$

$$\Upsilon_{k,t} := \mu_{k,t}^{\max} - \mu_{k,t}^{\min}, k \in \psi_n$$

$$h(\boldsymbol{x}, \boldsymbol{z}, \boldsymbol{\sigma}, \boldsymbol{\beta}, \boldsymbol{\Delta}) = \sum_{t=1}^{T}\sum_{k\in\psi_g}(c_{k0} - z_{l2}) - \sum_{k\in N}\left(\beta_k - \delta_k^{\min} + \delta_k^{\max}\right)g_k$$

$$+ \sum_{t=1}^{T}\{\sum_{k\in\psi_n}\{\Lambda_{k,t}P_{k,t}^d + H_{k,t}Q_{k,t}^d + \lambda_{k,t}^{\min}P_{gk}^{\min} - \lambda_{k,t}^{\max}P_{gk}^{\max}$$

$$+ \zeta_{k,t}^{\min}Q_{gk}^{\min} - \zeta_{k,t}^{\max}Q_{gk}^{\max} + \mu_{k,t}^{\min}\left(V_k^{\min}\right)^2 - \mu_{k,t}^{\max}\left(V_k^{\max}\right)^2\}$$

$$+ \sum_{k\in\psi_{es}}\{\rho_{k,t}^{\min}P_{ek}^{c,\min} - \rho_{k,t}^{\max}P_{ek}^{c,\max} + \theta_{k,t}^{\min}P_{ek}^{d,\min} - \theta_{k,t}^{\max}P_{ek}^{d,\max}$$

$$+ \xi_{k,t}^{\min}Q_{ek}^{\min} - \xi_{k,t}^{\max}Q_{ek}^{\max}\}\} + \sum_{t=1}^{T+1}\sum_{k\in\psi_{es}}\{\gamma_{k,t}^{\min}E_k^{\min} - \gamma_{k,t}^{\max}E_k^{\max}\}$$

$$\boldsymbol{A_t} = \sum_{k=1}^{n}\left(\Lambda_{k,t}\boldsymbol{Y}_k + H_{k,t}\overline{\boldsymbol{Y}}_k + \Upsilon_{k,t}\boldsymbol{M}_k\right)$$

In constraints (6.47) to (6.50) and (6.55), $t \in \{1, ..., T\}$, while $t \in \{2, ..., T\}$ for constraints (6.51). The decision variables include

$$\boldsymbol{x}_t := \left[\left(\boldsymbol{\lambda}_t^{\min}\right)^T, \left(\boldsymbol{\lambda}_t^{\max}\right)^T, \left(\boldsymbol{\zeta}_t^{\min}\right)^T, \left(\boldsymbol{\zeta}_t^{\max}\right)^T, \left(\boldsymbol{\mu}_t^{\min}\right)^T, \left(\boldsymbol{\mu}_t^{\max}\right)^T, \left(\boldsymbol{\gamma}_t^{\min}\right)^T, \left(\boldsymbol{\gamma}_t^{\max}\right)^T, \right.$$
$$\left. \left(\boldsymbol{\rho}_t^{\min}\right)^T, \left(\boldsymbol{\rho}_t^{\max}\right)^T, \left(\boldsymbol{\theta}_t^{\min}\right)^T, \left(\boldsymbol{\theta}_t^{\max}\right)^T, \left(\boldsymbol{\xi}_t^{\min}\right)^T, \left(\boldsymbol{\xi}_t^{\max}\right)^T\right]^T$$

$$\boldsymbol{z}_{k,t} := \left[z_{l1,t}, z_{l2,t}\right]^T, \quad k \in \psi_g$$

$$\boldsymbol{\Delta} := \left[\left(\boldsymbol{\delta}^{\max}\right)^T, \left(\boldsymbol{\delta}^{\min}\right)^T, \boldsymbol{\delta}^T\right]^T$$

and the Lagrange multipliers σ, β corresponding to (6.32) to (6.44). Using the same procedure as derived in references [13] and [14], it can be proved that the formulation (6.46) to (6.55) is the dual of the OPF problem (6.31)–(6.44) and the duality gap is zero.

Condition 1 (necessary and sufficient condition): The dual problem has an optimal solution and the corresponding A_t^{opt} ($t = 1, ..., T$) has a zero eigenvalue of multiplicity 2.

Condition 2 (sufficient condition): The graph generated by the real part of the admittance matrix Re{Y} is strongly connected.

6.4.3 Optimal Solution of the OPF Problem

When the optimal solution of problem (6.46) to (6.55) exists, the corresponding optimal solution of the primary OPF problem can be obtained according to the following procedure.

1) For $t = 1, ..., T$, get the corresponding semidefinite matrix A_t^{opt} of $\left(x_t^{opt}, z_t^{opt}, \sigma_t^{opt}, \beta^{opt}, \Delta^{opt} \right)$.

2) After getting a non-zero vector $\left[U_{1t}^T, U_{2t}^T \right]^T$ in the null space of A_t^{opt}, then the optimal voltage value for the primary OPF problem can be calculated by

$$\dot{V}_t^{opt} = (\tau_{1t} + \tau_{2t}i)(U_{1t} + U_{2t}i) \tag{6.56}$$

The values of τ_{1t} and τ_{2t} can be computed using the KKT conditions and the known phase angle of the slack bus [14].

3) Verify that the solution $\dot{V}_t^{opt}$ satisfies the constraints of the primary OPF problem.

4) Get the nodal current and power injections of the OPF problem and calculate other concerned variables.

6.5 Simulation and Comparison

In this section, we will test the interior point-based method and the SDP-based method on two test systems. The first test system is the modified IEEE 9-bus system to simulate the connection of a pumped hydroelectric storage plant. The second test system is the modified IEEE 57-bus system to simulate the deployment of multiple battery energy storage systems. The parameters of the ESSs are shown in Table 6.1. The daily load profiles are built based on the historical load data of a real power system. The base power is 100 MVA. The resistances of some transformer branches are zero in the test systems. In order to satisfy the second condition of the SDP model [15], it is assumed that the resistances of these transformer branches are 10^{-5} p.u.

6.5.1 With a Single Energy Storage System

The single diagram of the IEEE 9-bus system is shown in Figure 6.2. A PHS plant is connected to bus 5. When the residue energy at the end of the cycle is forced to be equal to the initial energy, the optimization results are listed in Table 6.2. It can be seen that the production costs obtained by the IPM and the SDP-based method are almost the same.

Table 6.1 Parameters of the two ESSs.

Parameters	Unit	Values	
		Pumped hydro storage	Battery
$E_m^{\mathrm{max}}, E_m^{\mathrm{min}}$	MWh	1240, 124	40, 4
$E_{m,\,0}$	MWh	310	12
$P_{em}^{c,\,\mathrm{max}}, P_{em}^{c,\,\mathrm{min}}$	MW	100,0	10,0
$P_{em}^{d,\,\mathrm{max}}, P_{em}^{d,\,\mathrm{min}}$	MW	100,0	10,0
$Q_{em}^{\mathrm{max}}, Q_{em}^{\mathrm{min}}$	MVar	90, −90	10, −10
η_m^c	%	92	95
η_m^d	%	80	92

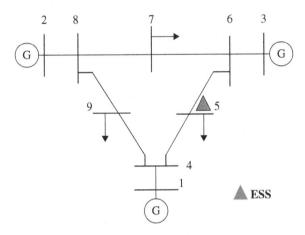

Figure 6.2 IEEE 9-bus test system with an ESS.

Table 6.2 Production costs of IEEE 9-Bus test system using different algorithms.

ESS	Production Cost ($)		Difference (%)
	SDP	Interior Point	
With ESS	1.3019×10^5	1.3022×10^5	0.02
Without ESS	1.3082×10^5	1.3082×10^5	0.00

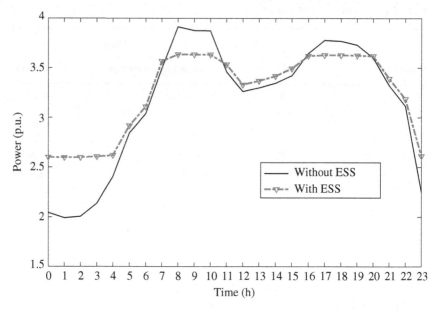

Figure 6.3 Total power output profiles of generating units with and without ESS.

Figure 6.3 illustrates the change of generators' total power output when the ESS participates in optimal operation of this system. The ESS discharges power during the time period of load peak and charges during the load trough. This charging/discharging schedule of ESS flattens the power output of the generators. In this case, the production cost is reduced slightly, which can be seen from Table 6.2.

The power and energy variations of the ESS within 24 hours are exhibited in Figure 6.4. Here, negative power means the ESS is discharging and positive power means the ESS is charging. In this figure, the area enclosed by the power curve and the abscissa is the energy that is absorbed (above the abscissa) and released (below the abscissa) by the ESS. The total energy absorbed in the whole day is about 236 MWh and the total released energy is about 174 MWh. The energy loss is about 62 MWh.

We then analyze the effectiveness or influence of the rolling optimization with an updated value of s_k. Before the OPF calculation of each time interval, the penalty coefficient s_k is recalculated according to

$$s_k = \alpha \frac{\Delta C}{E_{total}} \tag{6.57}$$

where ΔC is the production cost reduction when optimal dispatch of the ESS is considered, E_{total} is the total released energy of the ESS for OPF calculation of the last time interval, and α is a coefficient. In Figure 6.5, the residue energy curves

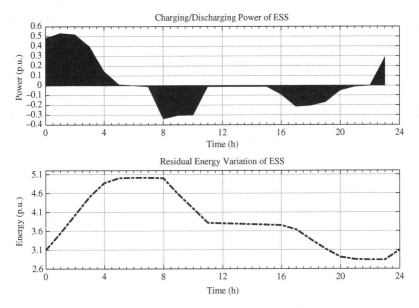

Figure 6.4 Active power and energy variation of ESS within 24 hours.

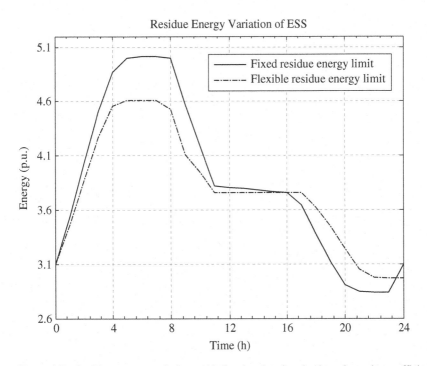

Figure 6.5 Residue energy variations with fixed and updated value of penalty coefficient.

of the ESS are illustrated and compared with fixed and updated s_k. When the value of residue energy of the ESS is considered in this case, the total energy absorption and release are reduced. However, the daily production cost is further decreased to about \$140 when compared to fixing the value of s_k.

6.5.2 With Multiple Energy Storage Systems

In this case, 10 battery storage systems are connected to the IEEE 57-bus system. The single line diagram of this test system is shown in Figure 6.6. The bus numbers connected with ESS are 4, 14, 20, 25, 30, 35, 40, 45, 50, and 55. The total capacity of the ESSs are 100 MW.

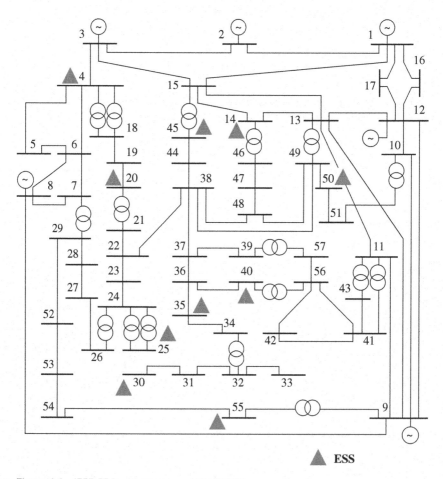

Figure 6.6 IEEE 57-bus test system with 10 ESSs.

Table 6.3 Production costs of IEEE 57-bus test system using different algorithms.

ESS	Production Cost ($)		Difference (%)
	SDP	Interior Point	
With ESS	7.3718×10^5	7.3726×10^5	0.01
Without ESS	7.4009×10^5	7.4022×10^5	0.02

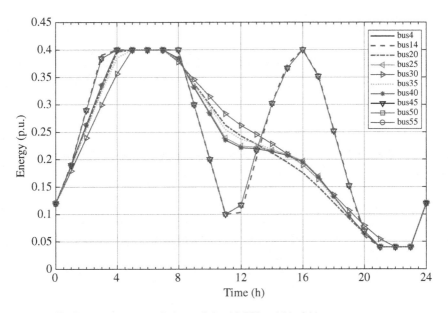

Figure 6.7 Residue energy variations of the 10 ESSs within 24 hours.

When it is enforced that the residual energy of each ESS is the same at the starting and ending time of the whole day, the optimization results of the IPM and SDP-based method are obtained and listed in Table 6.3. The differences in the production costs obtained using the two methods are very small. Figure 6.7 shows the residue energy variations of all 10 ESSs within the whole day. One can see that the energy variation of ESS is not only related to the time (load variation) but also to the location where the ESS is connected.

6.6 Conclusion and Discussion

For the researchers from the power system community and beyond, the AC OPF problem is a challenging and appealing one that has attracted a lot of research work. In this chapter, we focus on the OPF problem considering the optimal charging/discharging dispatch of energy-limited ESS. This problem is a multi-period optimization problem, which is different from the classical OPF problem without considering load variations. The basic OPF formulation with ESS and two different solution methods are introduced.

It is an active research topic on using convex relation and SDP-based techniques to solve the OPF problem, which will almost definitely lead to more advanced OPF solution methods. The following aspects are also very interesting and challenging, but have not been included in this chapter:

1) Power generation and load demand uncertainties.
2) Contingencies of transmission lines, transformers, etc.
3) Discrete variables.

References

1 Frank, S., Steponavice, I., and Rebennack, S. (2012). Optimal power flow: A bibliographic survey I. *Energy Systems* 3 (3): 221–258.

2 Frank, S., Steponavice, I., and Rebennack, S. (2012). Optimal power flow: A bibliographic survey II. *Energy Systems* 3 (3): 259–289.

3 Capitanescu, F., Glavic, M., Ernst, D. et al. (2007). Interior-point based algorithms for the solution of optimal power flow problems. *Electric Power Systems Research* 77 (5): 508–517.

4 Low, S.H. (2014). Convex relaxation of optimal power flow – Part I: Formulations and equivalence. *IEEE Transactions on Control of Network Systems* 1 (1): 15–27.

5 Low, S.H. (2014). Convex relaxation of optimal power flow – Part II: Exactness. *IEEE Transactions on Control of Network Systems* 1 (2): 177–189.

6 Chandy, K.M., Low, S.H., Topcu, U., and Xu, H. (2010). A simple optimal power flow model with energy storage. *49th IEEE Conference on Decision and Control (CDC)*, Atlanta, GA. IEEE, pp. 1051–1057.

7 Ding, H., Hu, Z., and Song, Y. (2015). Value of the energy storage system in an electric bus fast charging station. *Applied Energy* 157: 630–639.

8 Li, Z., Guo, Q., Sun, H. et al. (2015). Storage-like devices in load leveling: Complementarity constraints and a new and exact relaxation method. *Applied Energy* 151: 13–22.

9 Wang, X.F., Song, Y., and Irving, M. (2008). *Modern Power Systems Analysis*, 196–217. Springer Science and Business Media.

10 Baker, K., Zhu, D., Hug, G. et al. (2013). Jacobian singularities in optimal power flow problems caused by intertemporal constraints. *North American Power Symposium (NAPS)*: 1–6.

11 Baker, K., Guo, J., Hug, G. et al. (2016). Distributed MPC for efficient coordination of storage and renewable energy sources across control areas. *IEEE Transactions on Smart Grid* 7 (2): 992–1001.

12 Bai, X., Wei, H., Fujisawa, K. et al. (2008). Semidefinite programming for optimal power flow problems. *International Journal of Electrical Power and Energy Systems* 30 (6–7): 383–392.

13 Lavaei, J. and Low, S.H. (2012). Zero duality gap in optimal power flow problem. *IEEE Transactions on Power Systems* 27 (1): 92–107.

14 Gayme, D. and Topcu, U. (2013). Optimal power flow with large-scale storage integration. *IEEE Transactions on Power Systems* 28 (2): 709–717.

15 Lavaei, J. and Low, S.H. (2010). Convexification of optimal power flow problem. *2010 48th Annual Allerton Conference on Communication, Control, and Computing*. IEEE, pp. 223–232.

7

Power System Secondary Frequency Control with Fast Response Energy Storage System

7.1 Introduction

Balance between the power supply and demand should be maintained to achieve secure operation of power systems. The control of active power is usually equivalently called frequency control, which can be divided into three different timescales. They are named as primary frequency control, secondary frequency control, and tertiary frequency control[1] [1, 2], respectively. Primary frequency control is triggered when the frequency deviation is beyond the pre-set dead zone, which aims at stabilizing the system frequency as fast as possible. It is committed by all qualified generating units and executed by the turbine speed controllers of the units. As the primary frequency control is a proportional regulation and typically sustains up to tens of seconds, it is usually not sufficient to restore the system frequency and secondary frequency control (SFC) should be employed. We also call SFC automatic generation control (AGC) in this chapter. The power system operator is responsible for the operation of SFC or AGC. For an interconnected power system, maintaining frequency, and power interchanges with neighboring control areas at the scheduled values are the two main objectives of SFC [3]. With regards to tertiary control, it is actually the economic dispatch problem to balance power supply and demand at a timescale from several minutes to several days.

The SFC is a classical problem for power system analysis and control, which has attracted a lot of research. Today, the SFC or AGC is almost a must-have application for the energy management system (EMS) deployed at a power system control center. With the large-scale integration of renewable energy generation (REG), the fluctuation of REG brings new challenges for SFC. In addition, much research work has been done on SFC in considering the participation of new types of resources including energy storage system, electric vehicle, and other flexible

---

1 In some literatures, they are simply called primary control, secondary control, and tertiary control.

Energy Storage for Power System Planning and Operation, First Edition. Zechun Hu.
© 2020 John Wiley & Sons Singapore Pte. Ltd.
Published 2020 by John Wiley & Sons Singapore Pte. Ltd.

demand-side resources. With the progress of energy storage technologies, energy storage systems (i.e. battery storage and flywheel) have been successfully served as frequency regulation providers. These storage systems can respond to the regulation signals sent from the control center at significantly higher speeds compared to conventional generators (CGs). In order to provide an incentive for the fast responding resources to participate in SFC, the Federal Energy Regulatory Commission (FERC) of the United States enacted FERC Order 755, which requires system operators to add a performance payment to the qualified resources participating in the ancillary service of SFC [4]. Now, the performance payment scheme has already been implemented by almost all the independent system operators and regional transmission organizations in the United States.

Except for the payment scheme, the control strategy for the fast response energy storage system is another important issue. The fast response capability of battery ESS is considered and a new area control error (ACE) allocation strategy is proposed to improve the SFC performance in reference [5]. In reference [6], a method of allocating AGC signals to a set of generator and energy storage systems with different ramping, energy, and power limits is proposed. The method can maintain the high availability of energy-limited ESSs and improve the frequency regulation performance. The ESS can also be installed in a traditional power plant to participate in SFC along with thermal units in a coordinated fashion. Reference [7] introduces a real case of this kind of application.

In this chapter, a simulation method for SFC is first introduced by considering the participation of ESS. Then a method to quantify the capacity requirement for SFC is presented based on the simulation method. The contributions of ESS in particular is analyzed. Several control strategies for both conventional generating unit and ESS are compared in order to improve the frequency control performance by making better use of ESS.

7.2 Simulation of SFC with the Participation of Energy Storage System

We first consider a single-area system with one equivalent CG and one ESS. For the sake of simplicity, we assume that a regulation resource provides the same capacity for regulation up and regulation down services and that the ESS has the same charge and discharge power capacities.

7.2.1 Overview of SFC for a Single-Area System

In many studies, simulations are carried out to verify different AGC control strategies. A commonly used block diagram is illustrated in Figure 7.1 [3, 8–10]. In this figure, T_g and T_t are the governor and turbine time constants of the generating

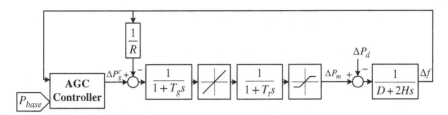

Figure 7.1 Block diagram of frequency control for a single-area system.

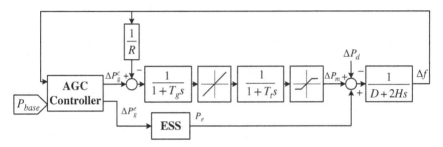

Figure 7.2 Block diagram of frequency control with generating unit and ESS.

unit, respectively. This model actually stands for a non-reheat generating unit [3, 8]. The primary frequency control of the generator is considered and R is called the speed regulation or droop, which is the ratio of frequency change to power output change [8], D is the load damping coefficient, H is the system inertia constant, ΔP_m and ΔP_d are the turbine's mechanical power change and load change, respectively, and Δf is the frequency deviation.

In this chapter, we will consider the situation where ESS participates SFC along with the generating unit, as shown in Figure 7.2. The AGC controller will allocate the control orders ΔP_g^c and ΔP_e^c to the generating unit and the ESS, respectively, and P_e is the power exchange of ESS with the power system. It is assumed that the base power of ESS for SFC is zero.

It should be noted that many published papers assumed a load disturbance and tested the frequency response after the disturbance for a time duration between seconds to minutes [11, 12]. The simulations in reference [10] used 40-minute test signals of AGC. With the increase in the considered time span to assess the frequency control performance, the required calculation time for the simulations will increase. We will evaluate the AGC performance and quantify the capacity requirement for SFC within a whole day. In order to reduce the simulation time, a fast simulation method based on numerical calculation will be given in the following sections.

7.2.2 Modeling of CG and ESS as Regulation Resources

In each AGC cycle, which is typically a few seconds in duration, the EMS sends control instructions to frequency regulation resources to adjust their power outputs. We use index $[k]$ for variables at the end of an AGC cycle k and $[0]$ for initial values (at the beginning of the first AGC cycle). Denote p_g, p_e as the power injected into the system by the CG and the ESS, respectively, and let p_d be the total power withdrawn by all the loads. Variable renewable energy generation, such as a wind farm can be regarded as a negative load. Note that the ESS can also withdraw power from the system and then p_e becomes negative.

A regulation resource has a base point that is decided by day-ahead and/or real-time economic dispatch. When providing the frequency regulation service, the regulation resource changes its power output around its base point according to the regulation signals sent from the dispatch center. The principal difference between a CG and an ESS in frequency regulation is how fast they can follow a regulation signal. The ESS changes its power almost instantaneously as long as the stored energy allows, while the CG has to ramp up (down) toward the target at a limited rate. Denote $\bar{r}_g$ (MW min^{-1}) as the ramp rate of the CG and $\bar{p}_e$ (MW) as the power capacity of the ESS. The regulation power contributed by the CG is

$$\Delta p_g[k] = p_g[k] - p_{g,base}[k] \tag{7.1}$$

where $p_{g,\,base}$ is the base point of the CG. As for the ESS, we assume that it provides only the regulation service so that the base point of p_e is always zero. Consequently, the ESS delivers a regulation power of $p_e[k]$.

Thus $\Delta p_g[k]$ and $p_e[k]$ are bounded by the regulation capacity $\Delta \bar{p}_g$ and $\bar{p}_e$, respectively. Regulation resources are required to reach their maximum regulation capacity within a specific duration τ (e.g. five minutes). This is usually not a problem for the ESS, but the regulation capacity of the CG has to be restricted by its ramping capability:

$$\Delta \bar{p}_g = \bar{r}_g \tau \tag{7.2}$$

The system regulation capacity is then given by

$$\bar{p}_r = \Delta \bar{p}_g + \bar{p}_e = \bar{r}_g \tau + \bar{p}_e \tag{7.3}$$

which is the maximum regulation power from both regulation resources.

7.2.3 Calculation of System Frequency Deviation

Neglecting active power losses, the deviation of the system frequency Δf is caused by the power imbalance between generation and load in the system [1]:

$$2H \frac{d\Delta f(t)}{dt} = p_g(t) + p_e(t) - (p_d(t) + D\Delta f(t)) \tag{7.4}$$

where p_d is the load power. Note that the variables in Eq. (7.4) are continuous functions with regard to time t. Denote

$$\Delta p_d(t) = p_d(t) - p_{g,base}(t) \tag{7.5}$$

Then Eq. (7.4) turns into

$$2H \frac{d\Delta f(t)}{dt} + D\Delta f(t) = \Delta p_g(t) + p_e(t) - \Delta p_d(t) \tag{7.6}$$

Given the fact that an AGC cycle is fairly short, we assume that $\Delta p_g(t)$ and $\Delta p_d(t)$ change linearly within one AGC cycle, while $p_e(t)$ steps up or down instantaneously. For the AGC cycle k spanning from time $t = t_k$ to $t = t_k + \delta$, where δ is the duration of one AGC cycle, we have

$$p_g(t) = p_g(t_k) + r_g[k]t \tag{7.7}$$

$$p_d(t) = p_d(t_k) + r_d[k]t \tag{7.8}$$

$$p_e(t) = p_e[k-1] + (p_e[k] - p_e[k-1])u(t - t_k) \tag{7.9}$$

when $t_k \le t \le t_k + \delta$ and $r_g[k]$ and $r_d[k]$ are the changing rates of $\Delta p_g(t)$ and $\Delta p_d(t)$ in the AGC cycle k, respectively:

$$r_g[k] = \frac{\Delta p_g[k] - \Delta p_g[k-1]}{\delta} \tag{7.10}$$

$$r_d[k] = \frac{\Delta p_d[k] - \Delta p_d[k-1]}{\delta} \tag{7.11}$$

where $u(t)$ is the unit step function.

As a consequence, we obtain the analytic solution of Eq. (7.6) as

$$\Delta f(t) = (\Delta f(t_k) - a) \exp\left(-\frac{D}{2H}(t - t_k)\right) + \frac{r_g[k] - r_d[k]}{D}(t - t_k) + a \tag{7.12}$$

where

$$a = \frac{D\left(\Delta p_g[k-1] - \Delta p_d[k-1] + p_e[k]\right) - 2H\left(r_g[k] - r_d[k]\right)}{D^2}. \tag{7.13}$$

Therefore, we have

$$\Delta f[k] = \Delta f(t_k + \delta) = (\Delta f[k-1] - a) \exp\left(-\frac{D\delta}{2H}\right) + \frac{r_g[k] - r_d[k]}{D}\delta + a \tag{7.14}$$

Given the regulation power of the regulation resources (Δp_g and p_e) and the load variation (Δp_d), we can calculate $\Delta f[k]$ one by one via Eq. (7.14).

7.2.4 Estimation and Allocation of Regulation Power

The dispatch center controls the power output of each regulation resource to balance the generation and load in the power system. The process mainly involves two steps: first estimating the regulation power, i.e. the amount of power required to be injected into or withdrawn from the system to maintain the balance, and then allocating the regulation power to the regulation resources.

We estimate the mismatch between the generation and load in the system by the ACE:

$$ACE[k] = B\Delta f[k] \tag{7.15}$$

where B is the frequency bias factor of the system [3]. Since we are investigating a single-area system, ACE has only one item corresponding to the frequency deviation. The regulation power $\pi[k]$ is obtained by the ACE passed through a proportional-integral (PI) controller:

$$\pi[k] = -\omega_p \cdot ACE[k] - \omega_i \cdot IACE[k] \tag{7.16}$$

where ω_p, ω_i are the proportional and the integral gain, respectively, and $IACE[k]$ is the integral of ACE with a saturation limit $\overline{IACE}$:

$$IACE[k] = \min\left\{ \max\left\{ \sum_{i=1}^{k} ACE[i]\delta, -\overline{IACE} \right\}, +\overline{IACE} \right\} \tag{7.17}$$

The saturation limit for the integral item is rather helpful when the system is short of regulation capability; its absence may result in a large accumulation in IACE when the regulation capacity is used up, ending up with an overcorrection after the system frequency restores ($\Delta f \approx 0$).

The next step is to allocate the regulation power $\pi[k]$ between the CG and the ESS. In this chapter, we first consider that the regulation power is allocated to the CG ($\pi_g[k]$) and the ESS ($\pi_e[k]$) on a pro rata basis, which is common in practical AGC systems:

$$\pi_g[k] = \frac{\Delta \overline{p}_g}{\overline{p}_r} \pi[k] \tag{7.18}$$

$$\pi_e[k] = \frac{\overline{p}_e}{\overline{p}_r} \pi[k] \tag{7.19}$$

Note that the actual responses of the regulation sources ($\Delta p_g[k]$ and $p_e[k]$) to the regulation signals ($\pi_g[k]$ and $\pi_e[k]$) are constrained by their physical characteristics. Specifically, for the CG, we have

$$\Delta p_g[k] = \begin{cases} \min\left\{ \pi_g[k], \Delta \overline{p}_g, \Delta p_g[k-1] + \overline{r}_g \delta \right\}, & \pi_g[k] \geq 0 \\ \max\left\{ \pi_g[k], -\Delta \overline{p}_g, \Delta p_g[k-1] - \overline{r}_g \delta \right\}, & \pi_g[k] < 0 \end{cases} \tag{7.20}$$

With regard to the ESS, the stored energy at the end of the AGC cycle k is

$$e[k] = e[k-1] - p_e[k]\delta \tag{7.21}$$

which is limited by the energy capacity of the ESS:

$$\underline{e} \le e[k] \le \bar{e} \tag{7.22}$$

where $\bar{e}$ and $\underline{e}$ (MWh) are the maximum and the minimum stored energy in the ESS, respectively. Accordingly, the ESS delivers a regulation power of

$$p_e[k] = \begin{cases} \min\left\{\pi_e[k], \bar{p}_e, \dfrac{e[k-1]-\underline{e}}{\delta}\right\}, & \pi_e[k] \ge 0 \\[3mm] \max\left\{\pi_e[k], -\bar{p}_e, \dfrac{e[k-1]-\bar{e}}{\delta}\right\}, & \pi_e[k] < 0 \end{cases} \tag{7.23}$$

7.3 Capacity Requirement for Secondary Frequency Control with Energy Storage System

For a single-area system, the performance of frequency regulation Δf_σ is measured by the standard deviation of the system frequency excursion Δf. We define the regulation requirement κ as the minimum system regulation capacity $\bar{p}_r$, which ensures a specific level of Δf_σ:

$$\kappa = \min\{\bar{p}_r\} \tag{7.24}$$

subject to

$$\Delta f_\sigma \le \Delta \bar{f}_\sigma \tag{7.25}$$

where $\Delta \bar{f}_\sigma$ is the desired level of Δf_σ, e.g. 0.02 Hz.

7.3.1 Procedure to Quantify Regulation Capacity Requirements

The regulation performance Δf_σ can be calculated under the specific regulation capacities $\Delta \bar{p}_g$ and $\bar{p}_e$. Generally, the more frequency regulation capacity $\bar{p}_r$ there is, the better (smaller) the regulation performance Δf_σ can be achieved. In the light of this fact, we can find the regulation requirement κ via a bisection method.

We emphasize that the monotonicity of Δf_σ with regard to $\bar{p}_r$ cannot be ensured and is related to the estimation and allocation of regulation power. For example, an overestimated power imbalance and a large ESS in the power system may result in an overcorrection of Δf and consequently may cause deterioration of the regulation performance. However, a practical AGC system definitely comes with elaborately configured SFC parameters, so it is reasonable to assume that a larger $\bar{p}_r$ leads to a smaller Δf_σ. We will exemplify the monotonicity with numerical cases.

The ratio of $\bar{p}_e$ to $\bar{p}_r$ as η is fixed when we quantify the regulation capacity requirement. To start the iterative process, the initial value of $\bar{p}_r$, i.e. $\bar{p}_r^{(1)}$, can be arbitrarily set as long as $\bar{p}_r^{(1)} > 0$.

After the calculation of every $\Delta f[k]$, Δf_σ can be obtained. The $\bar{p}_r$ is decreased toward zero if $\Delta f_\sigma < \Delta \bar{f}_\sigma$ and increased otherwise. Then, we start the next run with the adjusted system regulation capacity $\bar{p}_r^{(n+1)}$. Specifically, after the nth run, we double the system regulation capacity in the next run, i.e.

$$\bar{p}_r^{(n+1)} = 2\bar{p}_r^{(n)} \tag{7.26}$$

when $\Delta f_\sigma > \Delta \bar{f}_\sigma$, and reduce $\bar{p}_r$ in the next run to the midpoint of $\bar{p}_r$ in the last two runs, i.e.

$$\bar{p}_r^{(n+1)} = \frac{\bar{p}_r^{(n)} + \bar{p}_r^{(n-1)}}{2} \tag{7.27}$$

when $\Delta f_\sigma < \Delta \bar{f}_\sigma$. The search stops when Δf_σ is close enough to $\Delta \bar{f}_\sigma$, i.e.

$$\Delta \bar{f}_\sigma - \varepsilon < \Delta f_\sigma < \Delta \bar{f}_\sigma \tag{7.28}$$

where ε is the preset tolerance.

The detailed flow chart of the calculation is presented in Figure 7.3. K is the number of AGC cycles in a day. Note that a proper setting of $\Delta \bar{f}_\sigma$ is necessary to obtain a desirable solution. If $\Delta \bar{f}_\sigma$ is too large, the system can maintain the balance by itself without any regulation capacity, i.e. $\kappa = 0$; if $\Delta \bar{f}_\sigma$ is too small, the system may fail to meet the objective no matter how much regulation capacity there is, i.e. $\kappa = +\infty$. Such extreme cases are not covered in Figure 7.3.

7.3.2 Case Studies

The load data in 15 days (including weekdays, weekends, and holidays) from a real power system is used to calculate the regulation requirements. The parameters used in the simulations are listed in Table 7.1. We tune these parameters according to the actual configuration of the AGC system.

The *energy-to-power ratio* of an ESS is defined as follows:

$$\xi = \frac{\bar{e} - \underline{e}}{\bar{p}_e} \tag{7.29}$$

which has a dimension of time. The typical ξ of a flywheel energy storage unit (ESS) is around 15 minutes and ξ of a battery energy storage system can be up to several hours. We first analyze the regulation requirements when $\xi = 15$ minutes.

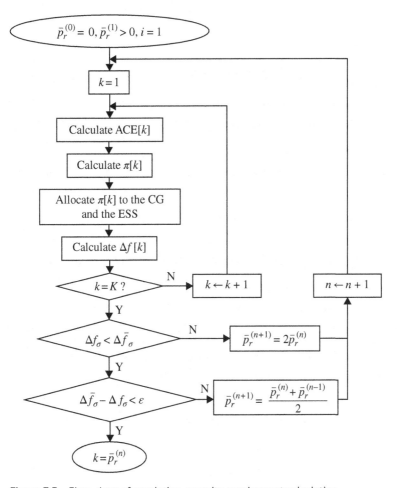

Figure 7.3 Flow chart of regulation capacity requirement calculation.

Table 7.1 Values of parameters used in simulations.

Parameter	Value
τ	5 min
δ	5 s
ε	10^{-4} Hz
D	2200 MW Hz^{-1}
H	11 000 MW $\cdot$ s Hz^{-1}
B	3400 MW Hz^{-1}
ω_p	1.0
ω_i	0.1 s^{-1}
$\overline{IACE}$	2000 MW $\cdot$ s

7.3.2.1 Regulation Capacity Requirements Under Different Regulation Resource Mixes

We first demonstrate the monotonicity of Δf_σ with regard to $\bar{p}_r$ so that the calculation method proposed above is applicable. The average Δf_σ of 15 days under five different resource mixes are exhibited in Figure 7.4. In each case, the regulation performance Δf_σ shows a continual but decelerating improvement as the system regulation capacity $\bar{p}_r$ increases. It can be seen that the trend is consistent for all 15 days simulated.

We next calculate the regulation requirement κ using the proposed approach. The average κ of 15 days obtained under different η and $\Delta \overline{f}_\sigma$ values are shown in Figure 7.5. The results indicate that the system requires more regulation capacity when a better regulation performance $\Delta \overline{f}_\sigma$ is desired. When the ESS constitutes a small portion of the regulation capacity, κ drops as η increases, but when η continues to rise, κ begins to climb and picks up speed. The reason is that the ESS undertakes more regulation power under a higher η and becomes more likely to get fully charged (discharged), unable to respond to the subsequent regulation down (up) signals. As a consequence, a higher energy capacity of the ESS is required to improve its availability and the demand for the power capacity of ESS rises proportionally because ξ is fixed.

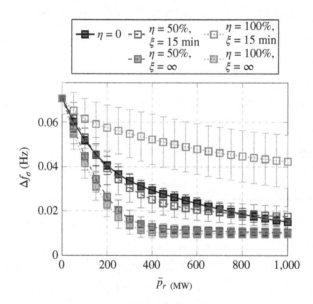

Figure 7.4 Relationship between regulation performance (Δf_σ) and capacity ($\bar{p}_r$) under different resource mixes ($\bar{p}_r$) and ESS energy capacities (ξ).

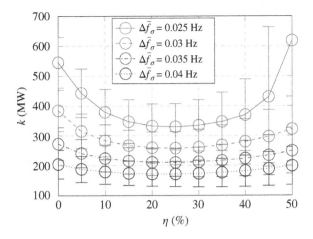

Figure 7.5 Regulation capacity requirements (κ) under different regulation resource mixes (η) and desired performance ($\Delta \bar{f}_\sigma$), ξ = 15 minutes.

7.3.2.2 Relative Effectiveness of ESS to CG for Frequency Regulation

We have shown that an appropriate proportion of ESS reduces regulation requirements. From another perspective, we can consider that the ESS is a more effective regulation resource compared to the CG. If ϑ MW of the regulation capacity from a CG can be replaced by 1 MW from an ESS, while keeping the same regulation performance Δf_σ, we define ϑ as the *relative effectiveness* of the ESS to the CG. When the proportion of the ESS grows from η_1 to η_2, the regulation requirement changes from κ_1 to κ_2; then ϑ is given by the ratio of the reduced regulation capacity from the CG to the increased regulation capacity from the ESS:

$$\vartheta = \frac{(1 - \eta_1)\kappa_1 - (1 - \eta_2)\kappa_2}{\eta_2 \kappa_2 - \eta_1 \kappa_1} \tag{7.30}$$

The results under different $\Delta \bar{f}_\sigma$ values are shown in Figure 7.6. When η is fairly small, each MW from the ESS can replace around 3 to 6 MW from the CG on average; the stricter $\Delta \bar{f}_\sigma$ is, the more effective the ESS becomes. As η increases, a quickly diminishing ϑ is observed; when $\eta > 25\%$, the ESS is less effective than the CG ($\vartheta < 1$).

Considering the relatively high cost of ESSs, it is not economical to add more ESSs when they already constitute a large portion of the total regulation capacity. Consequently, an appropriate mix of CGs and ESSs is preferable over using a single type of regulation resource in the system.

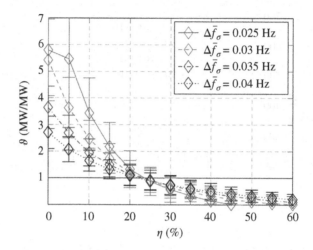

Figure 7.6 Relative effectiveness of ESS to CG (ϑ) as a function of ESS percentage (η) under different performance targets ($\Delta \bar{f}_\sigma$), ξ = 15 minutes.

The two types of regulation resources considered here correspond to two extreme cases: CGs have the slowest ramp rate allowed by the grid operator, while ESSs have the fastest ramping rate. For the two specific regulation resources, the fast-ramping one is more effective than the slow one, and the upper limit of the relative effectiveness is given by ϑ; the result does not necessarily mean that an arbitrary ESS is superior to a CG. For example, a hydropower generator (CG) and a compressed air ESS may have similar ramping capabilities.

7.3.2.3 Energy Capacity of ESS

The ESS cannot provide regulation up (down) service when its stored energy reaches the minimum (maximum) level; even worse, it has to withdraw its currently delivered regulation power immediately when the system still needs it, which deteriorates the regulation performance. As an example, we exhibit the regulation power from the CG and the ESS in a period of time in Figure 7.7. The ESS provides continuous regulation down power and becomes fully charged around $17:03, 17:10,$ and $17:15$ (the circles on Figure 7.7). It has to stop charging immediately, which subjects the system to a higher demand for regulation power with less regulation capacity available. In this situation, the sudden loss of the ESS's support is indeed putting more burden on frequency regulation of the CG.

We demonstrate the counterproductive behavior of the ESS by increasing η within a fixed $\bar{p}_r = 400$ MW. The regulation performance Δf_σ and the availability to provide a regulation service (both up and down) of the ESS under different ξ are presented in Figure 7.8. With a limited ξ, the availability of the ESS is generally

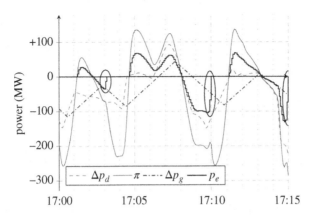

Figure 7.7 An example of counterproductive behavior of ESS during the regulation process ($\bar{p}_r$= 400 MW, η=50%, ξ= 15 minutes).

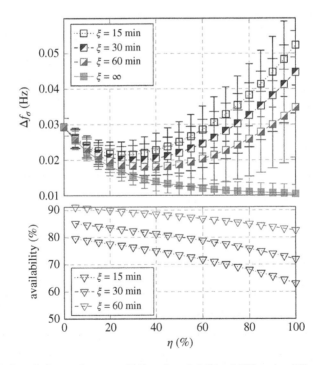

Figure 7.8 Regulation performance (Δf_σ) and availability of ESS under different proportions of ESS (η) with various energy capacities (ξ), $\bar{p}_r$ = 400 MW.

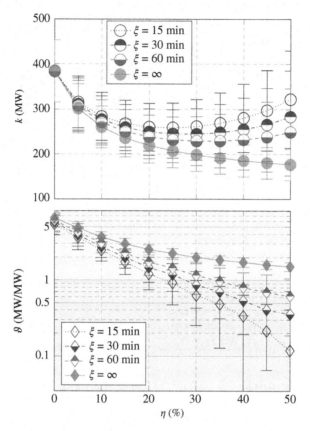

Figure 7.9 Regulation capacity requirements (κ) and relative effectiveness of ESS to CG (ϑ) under different proportions (η) and energy capacities (ξ) of ESS, $\Delta \bar{f}_\sigma$ = 0.03 Hz.

lower as η increases, neutralizing the benefits from the ESS and eventually worsening the regulation performance. The turning point appears at a larger η under a longer ξ.

The regulation requirements and the relative effectiveness of the ESS to the CG under different ξ are exhibited in Figure 7.9. It can be seen that the shorter ξ is, the lower the availability of the ESS becomes due to its limited energy capacity and κ goes up faster as η grows. With regard to the relative effectiveness, the ESS with a limited energy capacity outperforms the CG only when it makes up a small proportion (e.g. less than 30%) of the system regulation capacity, while ϑ never drops below 1 when the ESS has unlimited energy. This implies that a minimum requirement of ξ and a cap for η may be necessary to ensure the availability of a regulation service provided by an ESS.

7.4 Control Strategies of Secondary Frequency Control with Energy Storage System

7.4.1 CG First Power Allocation Strategy

In the above-mentioned simulations, the regulation power $\pi[k]$ is allocated to the CG and the ESS on a pro rata basis, which may not make full use of the fast response capability of ESS. The AGC system allocates $\pi_g[k]$ to the CG and $\pi_e[k]$ to the ESS, while the fulfilled regulation powers are $\Delta p_g[k]$ and $p_s[k]$, respectively. Now we want to find an allocation strategy that minimizes the unfulfilled regulation power:

$$\min f\left(\pi_g[k], \pi_s[k]\right) = \min_{\pi_g[k], \pi_s[k]} \left| \pi[k] - \left(\Delta p_g[k] + p_s[k] \right) \right| \tag{7.31}$$

subject to

$$\pi_g[k] + \pi_s[k] = \pi[k] \tag{7.32}$$

We define

$$\Delta p_g^M[k] = \min\left\{ \Delta \bar{p}_g, \Delta p_g[k-1] + \bar{r}_g \delta \right\} \tag{7.33}$$

$$\Delta p_g^m[k] = \max\left\{ -\Delta \bar{p}_g, \Delta p_g[k-1] - \bar{r}_g \delta \right\} \tag{7.34}$$

as the maximum and minimum limits of $\Delta p_g[k]$ that can be adjusted within the AGC cycle k. Similarly,

$$p_s^M[k] = \min\left\{ \bar{p}_s, \frac{e[k-1] - e^m}{\delta} \right\} \tag{7.35}$$

$$p_s^m[k] = \max\left\{ -\bar{p}_s, \frac{e[k-1] - e^M}{\delta} \right\} \tag{7.36}$$

are defined as the maximum and minimum limits of $p_s[k]$ that can be adjusted within the AGC cycle k. Now Eqs. (7.20) and (7.23) can be respectively rewritten as

$$\Delta p_g[k] = \min\left\{ \max\left\{ \pi_g[k], \Delta p_g^m[k] \right\}, \Delta p_g^M[k] \right\} \tag{7.37}$$

$$p_s[k] = \min\left\{ \max\left\{ \pi_s[k], p_s^m[k] \right\}, p_s^M[k] \right\} \tag{7.38}$$

If $\pi[k] > \Delta p_g^M[k] + p_s^M[k]$, it means that the two regulation resources cannot satisfy the regulation power requirement. In this situation, the regulation power allocated to the CG and the ESS should meet the following conditions:

$$\pi_g[k] \geq \Delta p_g^M[k] \tag{7.39}$$

$$\pi_s[k] \geq p_s^M[k] \tag{7.40}$$

Then the minimal value of the objective function is

$$\min f\left(\pi_g[k], \pi_s[k]\right) = \pi[k] - \left(\Delta p_g^M[k] + p_s^M[k]\right) \tag{7.41}$$

In the same way, when $\pi[k] < \Delta p_g^m[k] + p_s^m[k]$, the allocated power should satisfy

$$\pi_g[k] \leq \Delta p_g^m[k] \tag{7.42}$$

$$\pi_s[k] \leq p_s^m[k] \tag{7.43}$$

The minimal value of the objective function is

$$\min f\left(\pi_g[k], \pi_s[k]\right) = \left(\Delta p_g^m[k] + p_s^m[k]\right) - \pi[k] \tag{7.44}$$

When $\pi[k]$ meets

$$\Delta p_g^m[k] + p_s^m[k] \leq \pi[k] \leq \Delta p_g^M[k] + p_s^M[k] \tag{7.45}$$

an appropriate allocation method can be designed to make

$$\Delta p_g[k] + p_s[k] = \pi[k] \tag{7.46}$$

and then the value of the objective function is zero. It can easily be proved that the sufficient and necessary condition for Eq. (7.46) is

$$\begin{cases} \Delta p_g^m[k] \leq \pi_g[k] \leq \Delta p_g^M[k] \\ p_s^m[k] \leq \pi_s[k] \leq p_s^M[k] \end{cases} \tag{7.47}$$

From Eqs. (7.37) and (7.38), the equivalent condition can be obtained:

$$\begin{cases} \Delta p_g[k] = \pi_g[k] \\ p_s[k] = \pi_s[k] \end{cases} \tag{7.48}$$

The solution to satisfy (7.47) is not unique. Considering the limited energy of the ESS, it is desirable that the regulation burden of the ESS is minimized when (7.47) holds:

$$\min h\left(\pi_g[k], \pi_s[k]\right) = \min_{\pi_g[k], \pi_s[k]} |p_s[k]| \tag{7.49}$$

For $\pi[k]$ to satisfy (7.45), the optimal allocation strategy to minimize Eq. (7.49) is

$$\pi_g[k] = \begin{cases} \Delta p_g^m[k], & \left(\Delta p_g^m[k] + p_s^m[k] \leq \pi_k < \Delta p_g^m[k]\right) \\ \pi_k, & \Delta p_g^m[k] \leq \pi_k \leq \Delta p_g^M[k] \\ \Delta p_g^M[k], & \left(\Delta p_g^M[k] < \pi_k \leq \Delta p_g^M[k] + p_s^M[k]\right) \end{cases} \tag{7.50}$$

$$\pi_s[k] = \pi[k] - \pi_g[k] \tag{7.51}$$

The above two equations can be further expressed as

$$\begin{cases} \pi_g[k] = \min\left\{ \max\left\{ \pi[k], \Delta p_g^m[k] \right\}, \Delta p_g^M[k] \right\} \\ \pi_s[k] = \pi[k] - \pi_g[k] \end{cases} \tag{7.52}$$

This can be explained as follows: the regulation power $\pi[k]$ is allocated to the CG preferentially and the residual power unfulfilled by the CG will then be assigned to the ESS. In the following, we will call this strategy the "CG First" strategy.

7.4.2 Two Other Strategies

The first one is called the "ESS First" strategy. As the name implies, this strategy allocates the regulation power $\pi[k]$ to the ESS first and the residual power to the CG:

$$\begin{cases} \pi_s[k] = \min\left\{ \max\left\{ \pi[k], p_s^m[k] \right\}, p_s^M[k] \right\} \\ \pi_g[k] = \pi[k] - \pi_s[k] \end{cases} \tag{7.53}$$

It is evident that this strategy is opposite to the "CG First" strategy.

The second strategy is called the "Signal Decomposition" strategy. The regulation power $\pi[k]$ is decomposed through a first-order infinite impulse response filter (exponential smoothing filter) and its low-frequency component is assigned to the CG:

$$\pi_g[k] = \beta \cdot \pi[k] + (1 - \beta) \cdot \pi_g[k-1] \tag{7.54}$$

while the high-frequency component is allocated to the ESS:

$$\pi_s[k] = \pi[k] - \pi_g[k] \tag{7.55}$$

The β in Eq. (7.54) is the smoothing factor. The filter's transfer function is

$$H(z) = \frac{\beta}{1 - (1-\beta)z^{-1}} \tag{7.56}$$

and the amplitude–frequency characteristic is

$$|H(j\omega)| = \left| \frac{\beta}{\sqrt{1 - 2(1-\beta)\cos\omega + (1-\beta)^2}} \right| \tag{7.57}$$

If the cut-off frequency is f_c (cut-off time T_c), then

$$\omega_c = 2\pi \frac{f_c}{f_s} = 2\pi \frac{T_s}{T_c} \tag{7.58}$$

$$|H(j\omega_c)| = \left| \frac{\beta}{\sqrt{1 - 2(1-\beta)\cos\omega_c + (1-\beta)^2}} \right| = \frac{1}{\sqrt{2}} \tag{7.59}$$

Solving Eq. (7.59) we obtain

$$\beta = \cos\omega_c - 1 + \sqrt{\cos^2\omega_c - 4\cos\omega_c + 3} \tag{7.60}$$

The AGC system of PJM Interconnection adopts a similar strategy, where the regulation signal is decomposed into RegA and RegD signals [4]. It should be noted that $\pi[k]$ is obtained after the PI controller. Although this strategy decomposed the regulation power, the real-time capabilities of regulation resources are not considered.

7.4.3 Frequency Control Performance and Cost Comparisons

7.4.3.1 Two Indices to Evaluate Frequency Control Performance
The following two indices are used to compare the frequency control performance of different control strategies.

1) The mean value of the frequency excursion is

$$\Delta f_\mu = \frac{1}{K}\sum_{k=1}^{K}\Delta f[k] \tag{7.61}$$

This index is used to measure the average frequency deviation. The closer it is to zero, the better the control performance is.

2) The standard deviation of frequency excursion is

$$\Delta f_\sigma = \sqrt{\frac{1}{K-1}\sum_{k=1}^{K}\left(\Delta f[k] - \Delta f_\mu\right)^2} \tag{7.62}$$

Because the average value of ACE is commonly close to zero, the average values of the regulation powers $\pi[k]$ and the frequency excursion $\Delta f[k]$ are also very close to zero. Thus, the following index will be evaluated:

$$\Delta f_\sigma = \sqrt{\frac{1}{K}\sum_{k=1}^{K}(\Delta f[k])^2} \tag{7.63}$$

7.4.3.2 Simulation Results and Comparisons
The parameters are the same as in the case studies in Section 7.3.2. The cut-off time T_c is set as 60 seconds. In order to compare the frequency regulation performance

with different control strategies, the regulation power capacity is fixed at 400 MW and the ratio of ESS power η is changed from 0 to 100%. The results of Δf_σ are shown in Figure 7.10.

It can be seen from Figure 7.10 that the "CG First" strategy is better than the other strategies. The "ESS First" strategy outperforms the "Pro Rata" strategy only when η is smaller than 20%. It is beyond expectation that the "Signal Decomposition" strategy shows the worst performance.

Theoretically, the "ESS First" strategy allocates the regulation power to the ESS preferentially and this is equivalent to assigning the low-frequency component of regulation power to the ESS. By contrast, the "CG First" strategy distributes the residue regulation power that the CG cannot undertake to the ESS, which can use the merits of the two regulation resources. For both "Pro Rata" and "Signal Decomposition" strategies, the capabilities of the regulation resources may not be fully utilized. Although it looks like the "Signal Decomposition" strategy is more elaborately designed, it may be inferior to the simple "Pro Rata" strategy shown in Figure 7.10. We explain this phenomenon using an example illustrated in Figure 7.11. It can be seen that the ESS can follow the regulation signal promptly while it is confined by the maximum power capacity. The response of the CG to the regulation signal is limited by the ramping speed. There are two shortcomings for the "Signal Decomposition" strategy:

1) The regulation power is decomposed and assigned to the two types of regulation resources without considering their capabilities in real time. The allocated

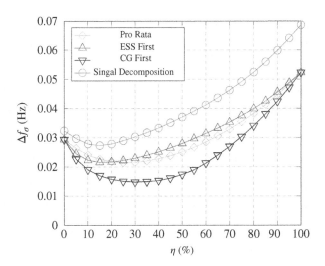

Figure 7.10 Control performances with the four control strategies ($\overline{p}_r$ = 400 MW, ξ=15 minutes).

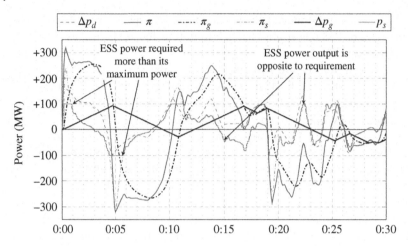

Figure 7.11 Regulation power allocation illustration with the signal decomposition strategy ($\bar{p}_r$ = 200 MW, η = 50%, ξ = 15 minutes, T_c = 300 seconds).

power may be larger than the maximum capacity of the ESS (e.g. at the time 0 : 01, 0 : 06 shown in Figure 7.11).

2) In some AGC cycles, the regulation power directions of the ESS and the CG are opposite. The power direction of the ESS is opposite to the regulation power $\pi[k]$ at time 0 : 15 and 0 : 22 in Figure 7.11.

In order to achieve a high-frequency control performance, the parameter β of the filter should be carefully selected while considering the mix of the regulation resources. However, the optimal value of β is system dependent and changes with the capacities of the regulation resources.

7.4.3.3 Frequency Control Costs

For the case study results shown in Figure 7.10, we see that the "CG First" strategy performs best and the "Signal Decomposition" strategy is not as good as expected. However, it should be noted that these performances are system dependent. Here we define three indices to further compare the frequency regulation costs with different control strategies.

1) Regulation energy
   This index is calculated as the integral of absolute regulation power over time and its unit is MWh:

$$w_g = \sum_{k=1}^{K} w_g[k] \qquad (7.64)$$

$$w_s = \sum_{k=1}^{K} |p_s[k]| \delta \tag{7.65}$$

where

$$w_g[k] = \begin{cases} \left| \dfrac{\Delta p_g[k-1] + \Delta p_g[k]}{2} \right| \delta & \left(\Delta p_g[k-1] \Delta p_g[k] > 0 \right) \\[4mm] \left| \dfrac{\left(\Delta p_g[k-1] \right)^2 + \left(\Delta p_g[k] \right)^2}{2 \left(\Delta p_g[k-1] - \Delta p_g[k] \right)} \right| \delta & \left(\Delta p_g[k-1] \Delta p_g[k] < 0 \right) \end{cases}$$

$$\tag{7.66}$$

The frequency control energy is actually the area enclosed by the real output curve and the baseline of a regulation resource. With a higher allocated regulation power for a longer duration, the frequency control energy will be larger. For the ESS, the baseline is kept at zero. However, Eq. (7.65) can be easily changed when the base power for the ESS is non-zero.

2) Regulation mileage
Regulation mileage is the sum of the absolute values of the frequency control signal movements [4]. It is defined as follows:

$$m_g = \sum_{k=1}^{K} \left| \Delta p_g[k] - \Delta p_g[k-1] \right| \tag{7.67}$$

$$m_s = \sum_{k=1}^{K} |p_s[k] - \Delta p_s[k-1]| \tag{7.68}$$

The unit of this index is ΔGW or ΔMW, which is different from that in reference [4]. The regulation mileage has been used in some power systems to calculate and compensate for the contribution of the regulation resource. It should be noted that calculation of the regulation mileage in a practical application should consider the regulation resource's real responses to the regulation signals.

3) Percentage of ESS availablility
The percentage of ESS availablility can be counted easily when the ESS provides the frequency control or regulation service. For the simulations, this index can be calculated by

$$\alpha = \frac{|\{k \mid e^m < e[k] < e^M\}|}{K} \times 100\% \tag{7.69}$$

This index can be used to reflect the effectiveness of a control strategy on managing the residue energy level for the ESS. A well-designed frequency control strategy should be able to maintain high availability of the ESS even when ξ is relatively small.

The average values for the 15 days are shown in Figures 7.12 and 7.13, with $\bar{p}_r = 400$ MW and $\xi = 15$ minutes. Comparing the four control strategies, it can be seen from the figures that:

1) With the "ESS First" strategy, the regulation energy and mileage of the CG decrease rapidly with the increase of η. The ESS takes much of the regulation burden and operates with a lower available percentage compared to other control strategies.
2) The regulation energy of the CG is similar with "CG First" and "Signal Decomposition" strategies. This indicates that the two control strategies make good use of the CG. However, with the "Signal Decomposition" strategy, the regulation energy and mileage decrease when η increases over 25%. This means that the capability of the ESS is not fully used. The phenomenon can also be seen from the index of ESS availability, which keeps almost 100% under the "Signal Decomposition" strategy. To achieve a better regulation performance, the parameter T_c should be tuned with the change of η.
3) For the "CG First" strategy, the ESS available percentage is more than 95% when η is below 50%. This verifies the fact that this strategy can properly manage an energy level of the ESS. It should also be noted that the energy requirement on the ESS will increase with the increase of η.
4) From Figures 7.10 and 7.12, we find that the total regulation energy and regulation mileage are positively related to the performance of the frequency regulation (Δf_σ). It is therefore easy to accept that generally more movements of regulation resources will lead to a better frequency control result.

7.5 Extending to Multi-area Power System

The simulation method described in Section 7.2 can be easily extended to a multi-area system with a gentle assumption. For an interconnected power system with multiple control areas, the ACE of each control area is calculated as [4]

$$ACE_i[k] = B_i \Delta f_i[k] + \Delta p_{tie,i}[k] \tag{7.70}$$

where $\Delta p_{tie,\,i}$ is the total tie-line power deviation of control area i. The performance index for frequency control should be changed from Δf_σ to the ACE standard deviation ΔACE_σ.

We performed simulations on a two-area interconnected power system. It is assumed that area 1 is the same as the above simulations and the regulation power

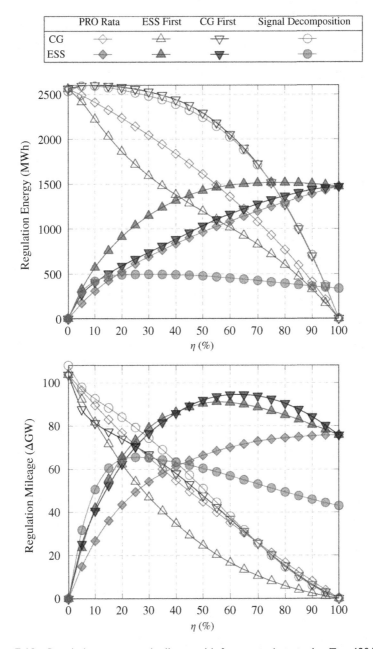

Figure 7.12 Regulation energy and mileage with four control strategies ($\overline{p}_r$ = 400 MW, ξ = 15 minutes).

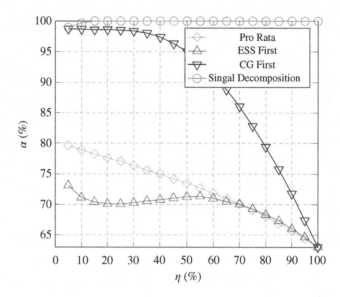

Figure 7.13 ESS availability percentage with four control strategies ($\bar{p}_r$ = 400 MW, ξ=15 minutes).

of area 2, $\bar{p}_{r,2}$, is 2000 MW ($\eta_2 = 0$). Keeping $\xi_1 = 15$ minutes and using the "CG First" control strategy, the frequency control performances for a whole day are illustrated in Figure 7.14.

In this figure, the colored surface graph represents the ACE standard deviation of area 1, $\Delta ACE_{1,\sigma}$, and the colored mesh graph stands for the ACE standard deviation of area 2, $\Delta ACE_{2,\sigma}$. The contour line of $\Delta ACE_{1,\sigma}$ is drawn at the top of Figure 7.14. The change of $\Delta ACE_{1,\sigma}$ is similar to the Δf_σ in the case studies of the single-area system, where $\Delta ACE_{1,\sigma}$ reduces with the increase in regulation power capacity and the proper ratio of the ESS. The variation of $\Delta ACE_{2,\sigma}$ is very small, which means that the impact of the regulation resource change of area 1 on the regulation performance of area 2 is negligible. Based on this result, the above discussions on the frequency control performance with different control strategies are valid for multi-area interconnected power systems.

7.6 Conclusion and Discussion

In order to simulate the SFC process in a longer time (e.g. a whole day) at high speed, this chapter first gives a simulation method with a simplified model and regulation resources, i.e. one CG and one ESS. Then a method to quantify the

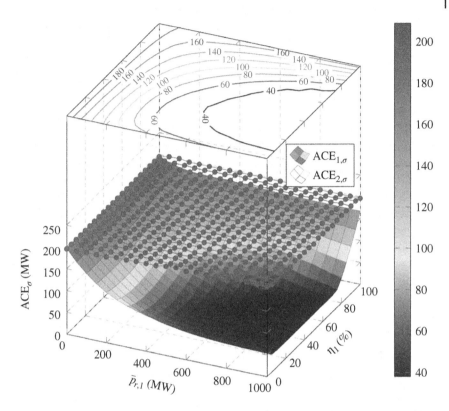

Figure 7.14 ACE standard deviations of a two-area interconnected power system with different $\bar{p}_{r,1}$ and η_1 of area 1 ($\bar{p}_{r,2}$ = 2000 MW, η_2 = 0, ξ_1 = 15 minutes, using the "CG First" strategy).

capacity requirement for SFC is introduced and case study results are given to show the role of ESS for SFC. Four strategies to control CG and ESS at the power control center are compared. It has been shown that the "CG First" strategy can achieve a better SFC performance than those of the other strategies. However, this conclusion may be system dependent. It is shown that the simulation method and control strategy for regulation resources can also be used in multi-area interconnected power systems. Last, there are two important points that should be discussed.

1) **SFC under the Control Performance Standard (CPS) criteria**
   The indices to evaluate the performance of SFC are different for different power systems. In the US, the North American Electric Reliability Council (NERC) adopted the CPS1 and CPS2 criteria to assess the SFC control performance from

1997 [13]. The formulation to calculate the capacity requirement for SFC under CPS criteria should be changed. The constraints are:

$$CPS1 \geq 100\% \tag{7.71}$$

$$CPS2 \geq 90\% \tag{7.72}$$

A procedure similar to that shown in Figure 7.3 can be used to evaluate the capacity requirement for SFC under CPS criteria.

2) **Control strategies for multiple frequency control resources**
   In the above analyses of this chapter, only one CG and one ESS are considered. In real power systems, there are normally multiple CGs and even a number of ESS participating SFCs. The four control strategies can be used in a real power system to determine $\pi_g[k]$ and $\pi_s[k]$ first. Then the simplest way to allocate the regulation power to each CG and each ESS is to use the proportional sharing principle.

References

1 Rebours, Y.G., Kirschen, D.S., Trotignon, M., and Rossignol, S. (February 2007). A survey of frequency and voltage control ancillary services – Part I: Technical features. *IEEE Transactions on Power Systems* 22 (1): 350–357.

2 Maurer, C., Krahl, S., and Weber, H. (2009). Dimensioning of secondary and tertiary control reserve by probabilistic methods. *European Transactions on Electrical Power* 19 (4): 544–552.

3 Bevrani, H. (2009). *Robust Power System Frequency Control*. New York: Springer.

4 Xu, B., Dvorkin, Y., Kirschen, D.S. et al. (2016). A comparison of policies on the participation of storage in US frequency regulation markets. *Power and Energy Society General Meeting (PESGM)*. IEEE, pp. 1–5.

5 Cheng, Y., Tabrizi, M., Sahni, M. et al. (2014). Dynamic available AGC based approach for enhancing utility scale energy storage performance. *IEEE Transactions on Smart Grid* 5 (2): 1070–1078.

6 Megel, O., Liu, T., Hill, D.J., and Andersson, G. (2018). Distributed secondary frequency control algorithm considering storage efficiency. *IEEE Transactions on Smart Grid* 9 (6): 6214–6228.

7 Xie, X., Guo, Y., Wang, B. et al. (2018). Improving AGC performance of coal-fueled thermal generators using multi-MW scale BESS: A practical application. *IEEE Transactions on Smart Grid* 9 (3): 1769–1777.

8 Kundur, P., Balu, N.J., and Lauby, M.G. (1994). *Power System Stability and Control*. New York: McGraw-Hill.

9 Wood, A.J., Wollenberg, B.F., and Sheblé, G.B. (2013). *Power Generation, Operation, and Control*. Wiley.

10 Luo, H., Hu, Z., Zhang, H. et al. (2018). Coordinated active power control strategy for deloaded wind turbines to improve regulation performance in AGC. *IEEE Transactions on Power Systems* 34 (1): 98–108.

11 Chakraborty, T., Watson, D., and Rodgers, M. (2017). Automatic generation control using an energy storage system in a wind park. *IEEE Transactions on Power Systems* 33 (1): 198–205.

12 Zhong, J., He, L., Li, C. et al. (2014). Coordinated control for large-scale EV charging facilities and energy storage devices participating in frequency regulation. *Applied Energy* 123: 253–262.

13 Gross, G. and Lee, J.W. (2016). Analysis of load frequency control performance assessment criteria. *IEEE Transactions on Power Systems* 16 (3): 520–525.

8

Integration of Large-Scale Energy Storage System into the Transmission Network

8.1 Introduction

Transmission expansion planning (TEP) is to decide where, when, and what type of new transmission lines should be built to supply the forecasted demand adequately [1]. A lot of research work on TEP has been carried out and published over the past few decades. TEP has drawn much attention from countries and regions with a rapid demand growth for electricity (e.g. China and Brazil) because of their imminent need for the expansion and reinforcement of transmission networks. In recent years, the increase in large-scale renewable energy generation (REG) connected to power grids has brought more attention to TEP [2].

Mathematically, TEP is a problem of finding the least costly plan of constructing new transmission facilities to meet power transmission requirements under certain security constraints [3, 4]. In general, the TEP problem is a mixed-integer, nonlinear and non-convex optimization problem that is quite difficult to solve [5]. To build a tractable formulation, the linear model based on DC power flow is widely adopted for simplicity. The high uncertainty over future electric demand and supply during the planning stage also makes the linear model acceptable for engineering application. Different types of methods have been proposed to solve the TEP problem, which can be classified into three types: heuristic, mathematical optimization, and artificial intelligence (AI) (e.g. genetic algorithm (GA), artificial neural network, and simulated annealing) [6]. The development of mature mathematical optimization algorithms and commercial software has widened the application of mixed-integer linear programming models and the corresponding optimization algorithm in TEP. Reference [7] proposed a disjunctive model for the transmission network expansion by introducing a big number M. Reference [8] proposed a method to determine the minimum value of M. The N-1 security criterion requires that the planned system should be able to withstand any single element outage of the network. The whole expansion planning problem becomes more complex

Energy Storage for Power System Planning and Operation, First Edition. Zechun Hu.
© 2020 John Wiley & Sons Singapore Pte. Ltd.
Published 2020 by John Wiley & Sons Singapore Pte. Ltd.

when this security criterion is considered. A mathematical model was built to deal with the transmission network expansion planning problem with N-1 security constraints in reference [9]. The upper-level model is solved using GA, while the lower-level model is solved by the linear programming method. More sophisticated methods for TEP considering N-1 contingencies and uncertainties can be found in references [10] and [11].

With the development of large-scale energy storage system (ESS) except for pumped hydroelectric storage system, e.g. a battery energy storage system, some research work [12–15] has been carried out on the optimal planning of ESS in transmission networks. In reference [12], a framework is proposed for the optimal placement of ESS based on the probabilistic optimal power flow formulation under an electricity market environment. A genetic algorithm is employed to search the optimal solutions to minimize the social cost and the benefit of increasing wind power utilization is also considered. Heuristic methods are implemented in references [12] and [13], while mathematical optimization methods are used in references [14] and [15]. In references [16] and [17], the transmission expansion planning and optimal ESS placement are considered simultaneously and the problems are solved by mathematical optimization methods.

In this chapter, the optimal integration of large-scale energy storage systems into the transmission network will be discussed. It should be noted that:

1) Optimal siting and sizing of the ESS problem is not considered alone. It is solved along with expansion planning of the transmission system.
2) Only a single-stage planning, i.e. static planning, is discussed in this chapter. The formulations can be expanded to multi-stage optimization without difficulty [18]. However, the computation burden will be increased.
3) It is assumed that the transmission network and ESS are centrally planned to maximize the social benefits.

8.2 Costs and Benefits of Investing ESS in a Transmission Network

Before we build the optimal planning for transmission and ESS expansion planning, we will take a look at the costs and benefits of investing ESS in a transmission network. Figure 8.1 illustrates the framework including the items of different costs, economic benefits, and social benefits [19]. Whereas the investment costs are mainly related to the type and capacity of an ESS, the benefits are more

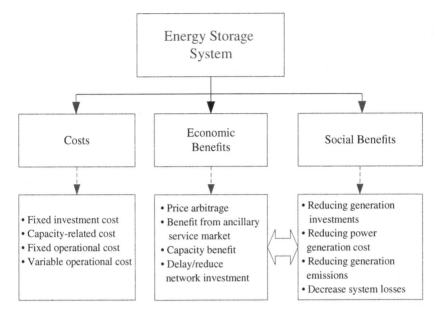

Figure 8.1 Costs and benefits of investing in an ESS.

complicated and closely associated with the power system operational conditions, power market, and other related policies.

The economic benefits listed in Figure 8.1 include price arbitrage, revenue from the participating ancillary market and capacity market, and deferral or reduction of network investment. In some power markets, such as PJM in the US, generation units can get payment from the capacity market. However, market rules on paying capacity costs to ESS are required. The benefit of delaying or reducing network investment costs is related to the regulations on generation and load connection to power systems. In some countries and regions, to connect a generator or a load, network investment costs should be fully or partially paid by the network user. Some incentives should be introduced for the ESS deployment, which reduce network reinforcement costs. Moreover, some benefits from building ESS are investor-related. Reduction in network investment costs and losses can be achieved if a transmission network company implement the investment, whereas the benefits of loss reduction can be considered as social benefits if the investment comes from other parties.

With respect to social benefits, the detailed analysis of social benefits brought by ESS can help governments in making subsidy policies for the construction of ESS. It is important because the ESS is typically expensive currently.

8.3 Transmission Expansion Planning Considering Energy Storage System and Active Power Loss

8.3.1 Objective Function and Constraints

It is assumed that ESSs are to be built and operated by the transmission service provider. The investment on transmission network expansion will be recovered with a reasonable rate of return, which reflects in the retail electricity price and eventually is paid by the network users. Therefore, the objective is to achieve the minimal equivalent annual cost F with the investment of transmission lines and ESSs, which is given as follows:

$$F(\mathbf{z}, \mathbf{q}, \mathbf{z}_e, x, W_e) = \tau_l F_l(\mathbf{z}) + F_q(\mathbf{q}) + \tau_e F_e(\mathbf{z}_e, \mathbf{p}_e, W_e) \tag{8.1}$$

where

$$F_l(\mathbf{z}) = \sum_{(i,j,k) \in \psi_{bn}} \gamma_{ij} z_{ij}^{(k)} \tag{8.2}$$

$$F_q(\mathbf{q}) = \sigma \sum_{(i,j,k) \in \psi_b} q_{ij}^{(k)} \tag{8.3}$$

$$F_e(\mathbf{z}_e, \mathbf{p}_e, W_e) = \sum_{i \in \psi_e} \left(\xi_{fi} z_{ei} + \xi_{pi} p_{ei} \right) + \xi_W W_e / \mu \tag{8.4}$$

represent the annualized costs of adding transmission lines, active power losses, and building ESSs, respectively. The investment costs (F_l and F_e) are converted to annual values by multiplying the capital recovery factors

$$\tau = \frac{r(1+r)^n}{(1+r)^n - 1}$$

where r is the discount rate and n is the lifespan in years, γ_{ij} stands for the capital cost of building a line in the corridor between node i and j, $z_{ij}^{(k)}$ is the binary decision variable for the candidate line (i, j, k), and k stands for the kth candidate line between nodes i and j.

The annual loss factor σ (in \$ MW^{-1}) in F_q (total loss cost) is the product of the retail electricity price and loss equivalent hours [16]. The total capital cost of ESS (F_e) consists of three parts: the fixed costs related to the decision variables (z_{ei}), the variable costs depending on the power capacities (p_{ei}), and the total energy capacity (W_e), while ξ_{fi} is the fixed cost of installing an ESS at node i, ξ_{pi} is the variable cost coefficient, and ξ_W is the per MWh cost of ESS. μ stands for the cycle efficiency of ESS.

Similar to the disjunctive TEP formulation [7], the following constraints can be easily derived:

$$p_{gi} + p_{ei} - \sum_{j,k} p_{ij}^{(k)} = p_{di}, \ \forall i \in \psi_n \tag{8.5}$$

$$W_e = W(P_e) \tag{8.6}$$

$$p_{ij}^{(k)} = f_{ij}^{(k)} + \frac{1}{2} q_{ij}^{(k)}, \forall (i,j,k) \in \psi_b \tag{8.7}$$

$$f_{ij}^{(k)} = z_{ij}^{(k)} \left(-b_{ij} \right) \sin \left(\delta_i - \delta_j \right), \ \forall (i,j,k) \in \psi_b \tag{8.8}$$

$$q_{ij}^{(k)} = 2z_{ij}^{(k)} g_{ij} \left[1 - \cos \left(\delta_i - \delta_j \right) \right], \forall (i,j,k) \in \psi_b \tag{8.9}$$

$$\left| p_{ij}^{(k)} \right| \le z_{ij}^{(k)} \overline{p}_{ij}, \ \forall (i,j,k) \in \psi_b \tag{8.10}$$

$$\underline{p}_{gi} \le p_{gi} \le \overline{p}_{gi}, \ \forall i \in \psi_g \tag{8.11}$$

$$z_{ei}\underline{p}_{ei} \le p_{ei} \le z_{ei}\overline{p}_{ei}, \ \forall i \in \psi_{es} \tag{8.12}$$

$$\delta_a = 0 \tag{8.13}$$

$$z_{ij}^{(k)} = 1, \ \forall (i,j,k) \in \psi_{b0} \tag{8.14}$$

$$z_{ij}^{(k)} = \{0,1\}, \ \forall (i,j,k) \in \psi_{bn} \tag{8.15}$$

$$z_{ei} = \{0,1\}, \ \forall i \in \psi_{es} \tag{8.16}$$

where ψ_n and ψ_g denote the sets of all buses and buses with generators, respectively; ψ_b, ψ_{b0}, and ψ_{bn} represent the sets of all (existing and candidate) lines, existing lines, and candidate lines, respectively; ψ_{es} is the set of candidate buses for installing ESS; p_{gi} and p_{ei} are the power injection of generator and energy storage to bus i, respectively; p_{di} represents the load demand at bus i; P_e is the total power of all ESSs; $p_{ij}^{(k)}$, $q_{ij}^{(k)}$, $f_{ij}^{(k)}$, and $z_{ij}^{(k)}$ denote the total active power, active power loss, active power without loss, and the decision variable of the kth line of the corridor (i, j); g_{ij}, b_{ij}, and $\overline{p}_{ij}$ are line conductance, susceptance, and maximum capacity in the corridor (i, j), respectively; δ_i, δ_j are the phase angle at bus i and j, respectively; node a is the slack bus; $\underline{p}_{gi}$ and $\overline{p}_{gi}$ are the lower and upper bounds of the generator power output at node i, while $\underline{p}_{ei}$ and $\overline{p}_{ei}$ are the lower and upper bounds of the ESS power capacity at node i, respectively.

The constraints (8.5) denote the power balance between power injection (from the generator and ESS, if any) and absorption (from transmission lines and demand) at each node. The constraints (8.6) present the relationship between power and energy capacity of ESSs, which will be discussed later. Power flow on line (i, j, k) is calculated by Eq. (8.7), each consisting of a lossless part given by Eq. (8.8) and a loss part given by Eq. (8.9). Line power flows, power outputs of generators, and ESSs are bounded by Eqs. (8.10) to (8.12) and the phase angle of the slack bus is set by Eq. (8.13).

It is assumed that the phase angle difference between two directly connected buses is small. This is the case under a normal condition and is commonly adopted in the published literature. Thus Eqs. (8.8) and (8.9) can be estimated by the first term of their Maclaurin series:

$$f_{ij}^{(k)} \simeq z_{ij}^{(k)}\left(-b_{ij}\right)\delta_{ij}, \forall(i,j,k) \in \psi_b \tag{8.17}$$

$$q_{ij}^{(k)} \simeq z_{ij}^{(k)}g_{ij}\delta_{ij}^2, \forall(i,j,k) \in \psi_b \tag{8.18}$$

A big number M is introduced to convert the above two equations into linear form as follows:

$$\left|f_{ij}^{(k)} + b_{ij}\delta_{ij}\right| \leq M\left(1 - z_{ij}^{(k)}\right), \ \forall(i,j,k) \in \psi_b \tag{8.19}$$

$$\left|q_{ij}^{(k)} - g_{ij}\delta_{ij}^2\right| \leq M\left(1 - z_{ij}^{(k)}\right), \ \forall(i,j,k) \in \psi_b \tag{8.20}$$

How to choose a proper value for M is discussed in reference [20]. There is a quadratic item in (8.20) that can be transformed into a linear expression by piecewise linearization.

8.3.2 Linearization of Line Losses

In reference [16], the advantages of the piecewise linearization approach with secant segments over that with chords is discussed. In Figure 8.2, the unknown variable δ_{ij} is divided into L ($L = 3$) segments within the given range $[0, D]$. The length of the lth segment is Δ_l and $\alpha_{ij}(l)$ and $\delta_{ij}(l)$ denote the slope and the value of segment l for the corridor (i, j), respectively.

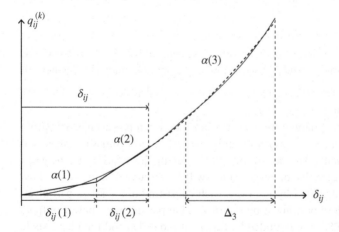

Figure 8.2 Piecewise linearization of line losses with secant segments ($L = 3$).

The active power loss in line (i, j, k) can be approximately calculated by

$$q_{ij}^{(k)} = g_{ij} \sum_{\ell \in \psi_\ell} \alpha(\ell) \delta_{ij}(\ell) \tag{8.21}$$

where ψ_ℓ is the set $\{1, 2, ..., L\}$. The following constraints should be satisfied:

$$\delta_{ij}(\ell) \leq \Delta_\ell \tag{8.22}$$

$$\delta_{ij} = \sum_{\ell \in \psi_\ell} \delta_{ij}(\ell) \tag{8.23}$$

It is proved in reference [16] that the optimal value of Δ_ℓ and $\alpha(\ell)$ are to achieve the minimal linear approximation error, as follows:

$$\Delta_1 = \frac{(1 + \sqrt{2})D}{1 + \sqrt{2}(2L - 1)} \tag{8.24}$$

$$\Delta_\ell = \frac{2\sqrt{2}D}{1 + \sqrt{2}(2L - 1)}, \forall \ell \in \psi_{\ell 2} \tag{8.25}$$

$$\alpha(1) = \frac{2D}{1 + \sqrt{2}(2L - 1)} \tag{8.26}$$

$$\alpha(\ell) = \frac{2 + 4\sqrt{2}(\ell - 1)}{1 + \sqrt{2}(2L - 1)}D, \forall \ell \in \psi_{\ell 2} \tag{8.27}$$

Here, $\psi_{\ell 2}$ is the set $\{2, ..., L\}$.

8.3.3 Sizing of Energy Storage Systems

For optimal placement of an ESS, both of its power capacity and energy capacity, should be determined. In the formulation discussed above, the energy capacity of each ESS in Eq. (8.6) is not explicitly given. To find the relationship between the total power capacity P_e and the total energy capacity W_e of each ESS, the load duration curve (LDC) [1] can be used. W_e should be no less than the capacity required in the worst case during the peak load day.

Figure 8.3 illustrates the requirement for W_e when an ESS is used to shave the peak load power by P_e and W_e is estimated using the rectangle method (midpoint approximation). The tail part of LDC (with a total length of $\max\{P_e\} = \sum_{i \in \psi_e} \overline{P}_{ei}$ in the p-axis) can be split into l blocks with equal length Δ_e and different heights $\beta(\ell)$, which should be determined from the LDC. Then W_e can be approximately expressed in a linear form as shown below:

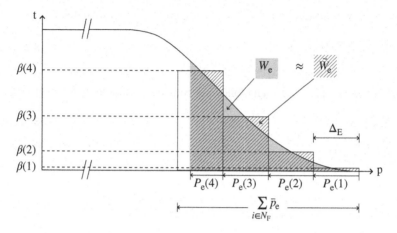

Figure 8.3 Representation of the energy capacity of an ESS by blocks ($l = 4$).

$$\int_{P_d^{max}-P_e}^{P_d^{max}} t\,dp = W_e \simeq \widetilde{W}_e = \sum_{\ell=1}^{l} \beta(\ell)P_e(\ell) \tag{8.28}$$

It should be noted that the allocation of energy capacity to each ESS cannot be decided by this method. It is also assumed that the ESSs can be fully charged during the off-peak hours. In peak hours, the load demand is supplied by both generating units and ESSs in the discharging state:

$$P_d = \sum_{i \in \psi_n} P_{di} = \sum_{i \in \psi_g} P_{gi} + \sum_{i \in \psi_{es}} P_{ei} \tag{8.29}$$

8.3.4 Complete Mathematical Formulation

According to the above derivations based on the disjunctive TEP formulation, the complete mathematical model for TEP considering optimal ESS placement is listed as follows:

$$\min \; \tau_1 \sum_{(i,j,k)\in\psi_{bn}} \gamma_{ij}z_{ij}^{(k)} + \sigma \sum_{(i,j,k)\in\psi_b} q_{ij}^{(k)} + \tau_e \left[\sum_{i\in\psi_{es}} \xi_{fi}z_{ei} + \sum_{i\in\psi_{es}} \xi_{pi}z_{ei} + \frac{\xi_w}{\mu} \sum_{\ell=1}^{l} \beta(\ell)P_e(\ell)\right] \tag{8.30}$$

s.t.

$$P_{gi} + P_{ei} - \sum_{j,k} p_{ij}^{(k)} = P_{di}, \; \forall i \in \psi_n \tag{8.31}$$

$$\sum_{i \in \psi_{es}} p_{ei} = \sum_{\ell = 1}^{l} P_e(\ell) \tag{8.32}$$

$$0 \leq P_e(\ell) \leq \frac{1}{l} \sum_{i \in \psi_{es}} \overline{p}_{ei}, \ \ell = 1, 2, ..., l \tag{8.33}$$

$$z_{ei} \underline{p}_{ei} \leq p_{ei} \leq z_{ei} \overline{p}_{ei}, \ \forall i \in \psi_{es} \tag{8.34}$$

$$p_{ij}^{(k)} = f_{ij}^{(k)} + \frac{1}{2} q_{ij}^{(k)}, \ \forall (i,j,k) \in \psi_b \tag{8.35}$$

$$f_{ij}^{(k)} = -b_{ij}\left(\delta_i - \delta_j\right), \ \forall (i,j,k) \in \psi_{b0} \tag{8.36}$$

$$\left| f_{ij}^{(k)} + b_{ij}\left(\delta_i - \delta_j\right) \right| \leq M\left(1 - z_{ij}^{(k)}\right), \ \forall (i,j,k) \in \psi_{bn} \tag{8.37}$$

$$q_{ij}^{(k)} = g_{ij} \sum_{\ell = 1}^{L} \alpha(\ell)\delta_{ij}(\ell), \ \forall (i,j,k) \in \psi_{b0} \tag{8.38}$$

$$0 \leq -q_{ij}^{(k)} + g_{ij} \sum_{\ell = 1}^{L} \alpha(\ell)\delta_{ij}(\ell) \leq M\left(1 - z_{ij}^{(k)}\right), \ \forall (i,j,k) \in \psi_{bn} \tag{8.39}$$

$$\left| p_{ij}^{(k)} \right| \leq \overline{p}_{ij}, \ \forall (i,j,k) \in \psi_{b0} \tag{8.40}$$

$$\left| p_{ij}^{(k)} \right| \leq z_{ij}^{(k)} \overline{p}_{ij}, \ \forall (i,j,k) \in \psi_{bn} \tag{8.41}$$

$$\left| f_{ij}^{(k)} \right| \leq \overline{p}_{ij}, \ \forall (i,j,k) \in \psi_{b0} \tag{8.42}$$

$$\left| f_{ij}^{(k)} \right| \leq z_{ij}^{(k)} \overline{p}_{ij}, \ \forall (i,j,k) \in \psi_{bn} \tag{8.43}$$

$$0 \leq q_{ij}^{(k)} \leq \overline{p}_{ij}, \ \forall (i,j,k) \in \psi_{b0} \tag{8.44}$$

$$0 \leq q_{ij}^{(k)} \leq z_{ij}^{(k)} \overline{p}_{ij}, \ \forall (i,j,k) \in \psi_{bn} \tag{8.45}$$

$$\underline{p}_{gi} \leq p_{gi} \leq \overline{p}_{gi}, \ \forall i \in \psi_g \tag{8.46}$$

$$\delta_{ij}^{+} + \delta_{ij}^{-} = \sum_{\ell = 1}^{L} \delta_{ij}(\ell), \ \forall (i,j) \in \psi_c \tag{8.47}$$

$$\delta_{ij}^{+} - \delta_{ij}^{-} = \delta_i - \delta_j, \ \forall (i,j) \in \psi_c \tag{8.48}$$

$$\delta_{ij}(\ell) \leq \Delta_\ell, \ \forall (i,j,k) \in \psi_{b0}, \ \forall \ell \in \psi_\ell \tag{8.49}$$

$$\delta_{ij}(\ell) \leq \Delta_\ell + M\left(1 - z_{ij}^{(k)}\right), \ \forall (i,j,k) \in \psi_{bn}, \ \forall \ell \in \psi_\ell \tag{8.50}$$

$$\delta_a = 0 \tag{8.51}$$

$$\delta_{ij}^{+} \geq 0, \ \delta_{ij}^{-} \geq 0, \ \forall (i,j) \in \psi_c \tag{8.52}$$

$$\delta_{ij}(\ell) \geq 0, \ \forall(i,j) \in \psi_c, \ \forall \ell \in \psi_\ell \tag{8.53}$$

$$\alpha(\ell) = \frac{2 + 4\sqrt{2}(\ell - 1)}{1 + \sqrt{2}(2L - 1)} D, \ \forall \ell \in \psi_\ell \tag{8.54}$$

$$\Delta_1 = \frac{(1 + \sqrt{2})D}{1 + \sqrt{2}(2L - 1)} \tag{8.55}$$

$$\Delta_\ell = \frac{2\sqrt{2}D}{1 + \sqrt{2}(2L - 1)}, \ \forall \ell \in \psi_{\ell 2} \tag{8.56}$$

$$z_{ij}^{(k)} = \{0, 1\}, \ \forall(i,j,k) \in \psi_{bn} \tag{8.57}$$

$$z_{ei} = \{0, 1\}, \ \forall i \in \psi_e \tag{8.58}$$

8.3.5 Case Studies

The above formulation is applied to the Garver 6-bus test system [21]. This test system has six nodes and six existing branches with a 760 MW load and 1110 MW generation capacity and γ_{ij} is set to \$1 000 000 per mile. For the cost of an ESS, $\xi_{pi} =$ \$500 kW^{-1}, $\xi_W =$ \$300 kWh^{-1}, $\mu = 75\%$, and $\xi_{fi} =$ \$100 000 are assumed. To convert capital costs to annual values, the discount rate r is assumed to be 10% and the life spans of each transmission line and each ESS are set to 30 and 15 years, respectively. The annual loss factor σ is approximately calculated according to the average retail electricity price in the USA (\$98.2 MWh^{-1}). The loss equivalent hours is set to 4897.9 hours per annum, which is estimated using the load profile given in reference [22]. The other parameters can be found in reference [16].

The solutions of three cases for the Garver 6-bus system are presented in Table 8.1 and Figure 8.4. It can be seen that when line losses are not considered, fewer new lines are required, but the systems will operate with higher losses, resulting in a higher total annual cost. This shows that taking line losses into consideration at the planning stage results in higher initial investments, while it is beneficial in the long run.

The results also indicate that with line losses considered, the application of ESSs reduces the transmission line investments. The deployment of ESS changes the power flow, deferring the upgrade for some heavily loaded lines or replacing required upgrades with less costly ones. However, it is noteworthy that the introduction of ESSs does not definitely lead to lower losses, but the optimal investment decision process ensures reduced overall annual costs. The cost reduction in this test case is not large, but ESSs have versatile applications in power system operation and can yield more revenue by providing ancillary services in off-peak days such as load following, spinning reserve, and frequency regulation.

Table 8.1 Results of different planning solutions for the Garver 6-bus system.

Items	Case 1 (no losses and no ESSs)	Case 2 (with losses and no ESSs)	Case 3 (with losses and ESSs)
Added lines (corridor and number of new lines)	3–5 (1) 4–6 (3)	2–6 (1) 3–5 (2) 4–6 (2)	2–3 (1) 2–6 (1) 3–5 (1) 4–6 (2)
Initial line investment, M$	110.00	160.00	130.00
Annual line cost, M$	11.67	16.97	13.79
Node of ESS and its power in MW	/	/	5 (12.50)
ESS capacity, MWh	/	/	12.17 (0.97 h)
Initial ESS investment, M$	/	/	7.77
Annual ESS cost, M$	/	/	1.02
Estimated loss, MW	/	30.97 (3.92%)	34.84 (4.45%)
Accurate loss, MW	55.56 (7.31%)	30.69 (3.88%)	34.60 (4.42%)
Estimation error, MW	–	+0.28	+0.20
Estimated annual loss cost, M$		14.90	16.76
Accurate annual loss cost, M$	26.72	14.76	16.64
Total initial investment, M$	110.00	160.00	137.77
Total annual cost, M$	38.39	31.74	31.45

The detailed simulation results on the IEEE 24-bus system can be found in reference [16].

8.4 Transmission Expansion Planning Considering Daily Operation of ESS

The formulation of TEP with ESS planning proposed in Section 8.3 only considers the operational constraints corresponding to the peak load level. The energy requirement for the ESS is estimated by the LDC. Thus, the method cannot take the varying operational costs of the whole system into account. In this section, the

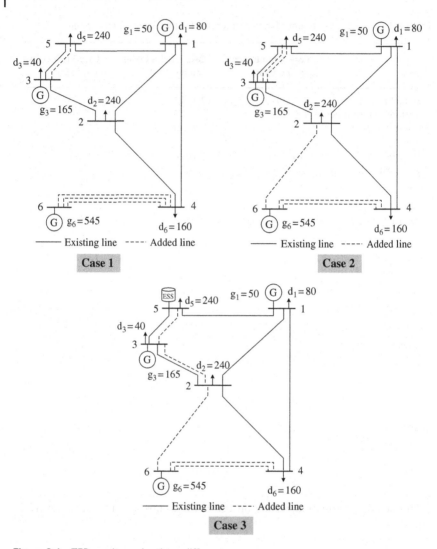

Figure 8.4 TEP results under three different cases.

formulation is extended to consider the daily operation of the system, including the optimal charging/discharging cycle of the ESS.

8.4.1 Different Approaches to Consider Optimal Daily Operation

Approaches that consider the optimal operation scheduling of generating units and ESSs can be divided into the following categories:

1) Robust optimization. This approach should also be based on daily operational optimization, where the energy balancing constraints of each ESS can be considered. The uncertainties of load consumption and renewable power generation are treated as uncertainty sets. In reference [17], an adaptive min–max–min cost model is introduced to find a robust optimal expansion plan for new lines and storages withstanding the worst-case realization of the uncertain variables. A decomposition algorithm using primal cutting planes is introduced to obtain the optimal solution. It should be noted that only a typical daily load profile and wind power profiles are dealt with. In addition, the lower and upper bounds of demand and wind power are considered for the robust optimization.
2) Scenario-based optimization. The typical scenarios or typical days are obtained to build the optimal planning formulation. The more scenarios considered, the more operational conditions including the charging/discharging schedule of ESS can be taken into account. However, the number of constraints and variables will also increase. In reference [14], a deterministic ESS investment optimization formulation is given first and then it is extended to a scenario-based optimization problem.
3) Simulation-based optimization. Although a scenario-based optimization method can deal with a number of different daily conditions, the number of total scenarios cannot be too large, e.g. more than 30, or the problem will become intractable, even for a medium-sized power system. To take into account detailed operational conditions, a simulation-based approach can be employed. Because many scenarios are considered in this approach, the TEP and ESS planning problem cannot be solved by mathematic programming methods. We need to turn to heuristic or trial and error methods for help. In reference [13], a three-stage planning procedure is proposed to get the near-optimal solution for the problem of siting and sizing of the ESS.

The pros and cons of the three approaches are summarized in Table 8.2. It also should be noted that the formulations proposed in references [13], [14], and [17] are all based on a DC power flow and active power losses are not considered.

8.4.2 Formulation of Scenario-Based Optimization

In this subsection, a formulation for TEP and ESS planning considering daily operation of the transmission system will be given. It is a scenario-based optimization approach to consider typical days within a whole year. Without wind or solar power generation, it is straightforward to get a number of typical days and the number of days that each typical day stands for using historical load data. It will be more complicated as large-scale wind farms and solar power plants are also involved.

Table 8.2 Pros and cons of different optimization approaches.

Approach	Pros	Cons
Robust optimization	Uncertainties of load and wind power are considered without requiring to know their probabilistic distributions.	The optimization problem is complicated and the formulation is based on one typical day, which may make the solution too conservative.
Scenario-based optimization	The variations of operational conditions can be properly considered.	The size of the optimization problem increases with the number of scenarios considered.
Simulation-based optimization	The operational conditions can be fully simulated and considered.	The computation burden is huge and optimality of the solution cannot be guaranteed by a heuristic method.

Assume ψ_s is the set of identified scenarios. The objective function of the planning problem is as follows:

$$
\min \quad \tau_1 \sum_{(i,j,k)\in\psi_{bn}} \gamma_{ij} z_{ij}^{(k)} + \sum_{s\in\psi_{sc}} \mu_s \sum_{t\in\psi_t} \left[\sum_{gi\in\psi_g} c_{gi} p_{gi,s,t} + \sum_{(i,j,k)\in\psi_b} q_{ij,s,t}^{(k)} \right]
$$
$$
+ \tau_e \left[\sum_{i\in\psi_{es}} \xi_{fi} z_{ei} + \sum_{i\in\psi_{es}} \xi_{pi} p_{ei} + \xi_{wi} w_{ei} \right] \tag{8.59}
$$

where ψ_t is the set of time intervals considered in each typical day (scenario), μ_s is the number of days that scenario s stands for and $\sum_{s\in\psi_{sc}} \mu_s = 365$ days. It should be noted that the energy capacity w_{ei} for a candidate ESS at node i is explicitly considered in the formulation.

For the investment of ESS, we can set the lower and upper bounds of power capacity and energy capacity and the energy to power ratio as follows:

$$
z_{ei}\underline{p}_{ei} \leq p_{ei} \leq z_{ei}\overline{p}_{ei}, \forall i \in \psi_{es} \tag{8.60}
$$

$$
z_{ei}\underline{W}_{ei} \leq W_{ei} \leq z_{ei}\overline{W}_{ei}, \forall i \in \psi_{es} \tag{8.61}
$$

$$
\underline{h}_{ei}p_{ei} \leq W_{ei} \leq \overline{h}_{ei}p_{ei}, \forall i \in \psi_{es} \tag{8.62}
$$

In (8.62), $\underline{h}_{ei}$ and $\overline{h}_{ei}$ are the lower and upper limits of the duration time in hours that the ESS can discharge at rated power p_{ei}.

For each typical day, the following constraints should be taken into account for each ESS i:

$$0 \leq p^c_{ei,s,t} \leq p_{ei}, \forall i \in \psi_{es}, s \in \psi_{sc}, t \in \psi_t \tag{8.63}$$

$$0 \leq p^d_{ei,s,t} \leq p_{ei}, \forall i \in \psi_{es}, s \in \psi_{sc}, t \in \psi_t \tag{8.64}$$

$$p_{ei,s,t} = p^d_{ei,s,t} - p^c_{ei,s,t}, \forall i \in \psi_{es}, s \in \psi_{sc}, t \in \psi_t \tag{8.65}$$

$$W_{ei,s,t} = W_{ei,s,t-1} + \left(\eta^c p^c_{ei,s,t} - \frac{p^d_{ei,s,t}}{\eta^d} \right) \Delta T, \forall i \in \psi_{es}, s \in \psi_{sc}, t \in \psi_t \tag{8.66}$$

$$\underline{soc}_{ei} W_{ei} \leq W_{ei,s,t} \leq \overline{soc}_{ei} W_{ei}, \forall i \in \psi_{es}, s \in \psi_{sc}, t \in \psi_t \tag{8.67}$$

Two variables $p^c_{ei,s,t}$ and $p^d_{ei,s,t}$ are introduced to identify the state of ESS. Because the charging efficiency (η^c) and discharging efficiency (η^d) are smaller than 1.0, the ESS will not charge ($p^c_{ei,s,t} > 0$) and discharge ($p^d_{ei,s,t} > 0$) at the same time in most of the cases. Here ΔT is the duration of each time interval and $\underline{soc}_{ei}$ and $\overline{soc}_{ei}$ are the lower and upper limits of the residue energy stored in the ESS.

It should be noted that the constraints for the generating units should also be considered:

$$\underline{P}_{gi} \leq P_{gi,s,t} \leq \overline{P}_{gi}, \quad \forall i \in \psi_g, s \in \psi_{sc}, t \in \psi_t \tag{8.68}$$

$$rd_{gi} \leq P_{gi,s,t} - P_{gi,s,t-1} \leq ru_{gi}, \quad \forall i \in \psi_g, s \in \psi_{sc}, t \in \psi_t \tag{8.69}$$

Constraints (8.68) limit the power output of each unit and constraints (8.69) represent the ramp-rate limits of each unit, where rd_{gi} and ru_{gi} stand for the ramp down and the ramp up limits of unit i, respectively.

The unit commitment is not considered here, readers can refer to Chapter 5 for details. For a real power system, generating units are scheduled to be in a maintenance state periodically. The available units that can be dispatched in different typical days may change. It is more practical to consider this factor.

For each scenario (typical day), constraints (8.31) to (8.53) will be changed into the following forms:

$$P_{gi,s,t} + P_{ei,s,t} + P_{wi,s,t} + P_{si,s,t} - \sum_{j,k} p^{(k)}_{ij,s,t} = P_{di,s,t}, \quad \forall i \in \psi_n, s \in \psi_{sc}, t \in \psi_t \tag{8.70}$$

$$p^{(k)}_{ij,s,t} = f^{(k)}_{ij,s,t} + \frac{1}{2} q^{(k)}_{ij,s,t}, \quad \forall (i,j,k) \in \psi_b, s \in \psi_{sc}, t \in \psi_t \tag{8.71}$$

$$f^{(k)}_{ij,s,t} = -b_{ij} \left(\delta_{i,s,t} - \delta_{j,s,t} \right), \quad \forall (i,j,k) \in \psi_{b0}, s \in \psi_{sc}, t \in \psi_t \tag{8.72}$$

$$\left| f_{ij,s,t}^{(k)} + b_{ij}\left(\delta_{i,s,t} - \delta_{j,s,t}\right) \right| \leq M\left(1 - z_{ij}^{(k)}\right), \ \forall (i,j,k) \in \psi_{bn}, s \in \psi_{sc}, t \in \psi_t$$

$$\text{(8.73)}$$

$$q_{ij,s,t}^{(k)} = g_{ij} \sum_{\ell=1}^{L} \alpha(\ell)\delta_{ij,s,t}(\ell), \ \forall (i,j,k) \in \psi_{b0}, s \in \psi_{sc}, t \in \psi_t \tag{8.74}$$

$$0 \leq -q_{ij,s,t}^{(k)} + g_{ij} \sum_{\ell=1}^{L} \alpha(\ell)\delta_{ij,s,t}(\ell) \leq M\left(1 - z_{ij}^{(k)}\right), \ \forall (i,j,k) \in \psi_{bn}, s \in \psi_{sc}, t \in \psi_t$$

$$\text{(8.75)}$$

$$\left| p_{ij,s,t}^{(k)} \right| \leq \overline{P}_{ij}, \ \forall (i,j,k) \in \psi_{b0}, s \in \psi_{sc}, t \in \psi_t \tag{8.76}$$

$$\left| p_{ij,s,t}^{(k)} \right| \leq z_{ij}^{(k)}\overline{P}_{ij}, \ \forall (i,j,k) \in \psi_{bn}, s \in \psi_{sc}, t \in \psi_t \tag{8.77}$$

$$\left| f_{ij,s,t}^{(k)} \right| \leq \overline{P}_{ij}, \ \forall (i,j,k) \in \psi_{b0}, s \in \psi_{sc}, t \in \psi_t \tag{8.78}$$

$$\left| f_{ij,s,t}^{(k)} \right| \leq z_{ij}^{(k)}\overline{P}_{ij}, \ \forall (i,j,k) \in \psi_{bn}, s \in \psi_{sc}, t \in \psi_t \tag{8.79}$$

$$0 \leq q_{ij,s,t}^{(k)} \leq \overline{P}_{ij}, \ \forall (i,j,k) \in \psi_{b0}, s \in \psi_{sc}, t \in \psi_t \tag{8.80}$$

$$0 \leq q_{ij,s,t}^{(k)} \leq z_{ij}^{(k)}\overline{P}_{ij}, \ \forall (i,j,k) \in \psi_{bn}, s \in \psi_{sc}, t \in \psi_t \tag{8.81}$$

$$\delta_{ij,s,t}^{+} + \delta_{ij,s,t}^{-} = \sum_{\ell=1}^{L} \delta_{ij,s,t}(\ell), \ \forall (i,j) \in \psi_c, s \in \psi_{sc}, t \in \psi_t \tag{8.82}$$

$$\delta_{ij,s,t}^{+} - \delta_{ij,s,t}^{-} = \delta_{i,s,t} - \delta_{j,s,t}, \ \forall (i,j) \in \psi_c, s \in \psi_{sc}, t \in \psi_t \tag{8.83}$$

$$\delta_{ij,s,t}(\ell) \leq \Delta_\ell, \ \forall (i,j,k) \in \psi_{b0}, \ , s \in \psi_{sc}, t \in \psi_t, \forall \ell \in \psi_\ell \tag{8.84}$$

$$\delta_{ij,s,t}(\ell) \leq \Delta_\ell + M\left(1 - z_{ij}^{(k)}\right), \ \forall (i,j,k) \in \psi_{bn}, s \in \psi_{sc}, t \in \psi_t, \ \forall \ell \in \psi_\ell$$

$$\text{(8.85)}$$

$$\delta_{a,s,t} = 0, s \in \psi_{sc}, t \in \psi_t \tag{8.86}$$

$$\delta_{ij,s,t}^{+} \geq 0, \ \delta_{ij,s,t}^{-} \geq 0, \ \forall (i,j) \in \psi_c, s \in \psi_{sc}, t \in \psi_t \tag{8.87}$$

$$\delta_{ij,s,t}(\ell) \geq 0, \ \forall (i,j) \in \psi_c, s \in \psi_{sc}, t \in \psi_t, \ \forall \ell \in \psi_\ell \tag{8.88}$$

The other constraints include (8.54) to (8.58), which are independent of scenarios.

It should be noted that wind power $p_{wi,s,t}$ and solar power $p_{si,s,t}$ are considered in constraint (8.70). Their outputs should be carefully quantified for each interval of each scenario before building this formulation. The whole formulation is a linear mixed-integer programming problem, which can be solved by off-the-shelf solvers. For a large-scale system, the size of the problem will become huge and it will take a long time to obtain a solution. A decomposition method [8] can be designed to solve the problem more efficiently.

8.5 Conclusion and Discussion

Traditional transmission expansion planning usually deals with the least cost network reinforcement scheme under the peak-load condition. By integrating the large-scale ESS into the transmission network, the network reinforcement cost may be reduced. However, since the energy capacity of ESS is limited, the optimal sizing of ESS should not only consider the power capacity but also the energy capacity of each ESS. In this chapter, a transmission expansion planning method considering ESS and active power loss is given in detail. The power loss of each line is linearized and the energy capacity requirement of each ESS is determined using the LDC. Simulation results show that the deployment of ESS can reduce the total annual cost. In Section 8.4, the TEP problem considering the daily operation of ESS is discussed. Different optimization methods to solve the problem are compared and the formulation of a scenario-based optimization is given.

References

1 Wang, X. and McDonald, J.R. (1994). *Modern Power System Planning*. New York, NY, USA: McGraw-Hill.

2 Orfanos, G.A., Georgilakis, P.S., and Hatziargyriou, N.D. (May 2013). Transmission expansion planning of systems with increasing wind power integration. *IEEE Transactions on Power Systems* 28 (2): 1355–1362.

3 Alguacil, N., Motto, A.L., and Conejo, A.J. (2003). Transmission expansion planning: a mixed-integer LP approach. *IEEE Transactions on Power Apparatus and Systems* 18 (3): 1070–1077.

4 Bahiense, L., Oliveira, G.C., Pereira, M., and Granville, S. (2001). A mixed integer disjunctive model for transmission network expansion. *IEEE Transactions on Power Apparatus and Systems* 16 (3): 560–565.

5 Zhang, H., Vittal, V., Heydt, G.T., and Quintero, J. (2012). A mixed-integer linear programming approach for multi-stage security-constrained transmission expansion planning. *IEEE Transactions on Power Apparatus and Systems* 27 (2): 1125–1133.

6 Latorre, G., Cruz, R.D., Areiza, J.M., and Villegas, A. (May 2003). Classification of publications and models on transmission expansion planning. *IEEE Transactions on Power Systems* 18 (2): 938–946.

7 Bahiense, L., Oliveira, G.C., Pereira, M., and Granville, S. (Aug. 2001). A mixed integer disjunctive model for transmission network expansion. *IEEE Transactions on Power Systems* 16 (3): 560–565.

8 Binato, S., Pereira, M.V.F., and Granville, S. (May 2001). A new Benders decomposition approach to solve power transmission network design problems. *IEEE Transactions on Power Systems* 16 (2): 235–240.

9 Silva, I.d.J., Rider, M.J., Romero, R. et al. (November 2005). Transmission network expansion planning with security constraints. *IEE Proceedings – Generation, Transmission and Distribution* 152 (6): 828–836.

10 Majidi-Qadikolai, M. and Baldick, R. (November 2016). Stochastic transmission capacity expansion planning with special scenario selection for integrating N-1 contingency analysis. *IEEE Transactions on Power Systems* 31 (6): 4901–4912.

11 Majidi-Qadikolai, M. and Baldick, R. (March 2018). A generalized decomposition framework for large-scale transmission expansion planning. *IEEE Transactions on Power Systems* 33 (2): 1635–1649.

12 Ghofrani, M., Arabali, A., Etezadi-Amoli, M. et al. (2013). A framework for optimal placement of energy storage units within a power system with high wind penetration. *IEEE Transactions on Sustainable Energy* 4 (2): 434–442.

13 Pandžić, H., Wang, Y., Qiu, T. et al. (2015). Near-optimal method for siting and sizing of distributed storage in a transmission network. *IEEE Transactions on Power Systems* 30 (5): 2288–2300.

14 Wogrin, S. and Gayme, D.F. (2015). Optimizing storage siting, sizing, and technology portfolios in transmission-constrained networks. *IEEE Transactions on Power Systems* 30 (6): 3304–3313.

15 Jabr, R.A., Džafić, I., and Pal, B.C. (2015). Robust optimization of storage investment on transmission networks. *IEEE Transactions on Power Systems* 30 (1): 531–539.

16 Zhang, F., Hu, Z., and Song, Y. (2013). Mixed-integer linear model for transmission expansion planning with line losses and energy storage systems. *IET Generation, Transmission and Distribution* 7 (8): 919–928.

17 Dehghan, S. and Amjady, N. (2016). Robust transmission and energy storage expansion planning in wind farm-integrated power systems considering transmission switching. *IEEE Transactions on Sustainable Energy* 7 (2): 765–774.

18 Vinasco, G., Rider, M.J., and Romero, R. (November 2011). A strategy to solve the multistage transmission expansion planning problem. *IEEE Transactions on Power Systems* 26 (4): 2574–2576.

19 Hu, Z., Zhang, F., and Li, B. (2012). Transmission expansion planning considering the deployment of energy storage systems. *2012 IEEE Power and Energy Society General Meeting.* IEEE, pp. 1–6.

20 Binato, S., Pereira, M.V.F., and Granville, S. (May 2001). A new Benders decomposition approach to solve power transmission network design problems. *IEEE Transactions on Power Systems* 16 (2): 235–240.

21 Garver, L.L. (1970). Transmission network estimation using linear programming. *IEEE Transactions on Power Apparatus and Systems* PAS-89 (7): 1688–1697.

22 Albrecht, P.F., Bhavaraju, M.P., Biggerstaff, B.E. et al. (1979). IEEE reliability test system. *IEEE Transactions on Power Apparatus and Systems* PAS-98 (6): 2047–2054.

9

Optimal Planning of the Distributed Energy Storage System

9.1 Introduction

Distributed generation (DG) has grown rapidly in recent years under the pressure of increased energy demand and environmental issues. High penetration of DGs can bring a significant impact on the existing distribution networks. The most serious problem is that the inherently high-volatility integrated DGs could threaten the secure and economic operation of distribution networks [1]. Integrating the distributed energy storage system (DESS) into the distribution network with DGs is an option to solve this problem. DESS can act as an effective tool to perform demand-side management, reduce the gap between load peaks and valleys, improve the utilization of electrical equipment and increase the penetration of renewable energy generation [2–4].

In recent years, many researchers have been studying the planning of DESS in the distribution network. Most researchers aim at optimizing the location, energy capacity, and power capacity of each DESS to minimize the total cost or to maximize the social benefits considering multiple factors, such as price arbitrage, active power loss reduction, uncertainty of DG power output, the life cycle of DESS, and so forth [5–10]. In reference [11], a convex optimization model is built for distribution network expansion planning integrating DESS. A method is proposed in reference [12] for the cost-effective improvement of system reliability through optimal planning of DESS in distribution systems. The methods on the allocation of DESS in power distribution networks are summarized in reference [13], including analytical methods, mathematical programming, and exhaustive search and heuristic methods. In this chapter, the optimal DESS planning problem for a distribution network with DGs will be introduced.

Energy Storage for Power System Planning and Operation, First Edition. Zechun Hu.
© 2020 John Wiley & Sons Singapore Pte. Ltd.
Published 2020 by John Wiley & Sons Singapore Pte. Ltd.

9.2 Benefits from Investing in DESS

For the investment of DESS in distribution networks, it is commonly considered that the distribution network operator (DNO) or utility company should be the investor. The benefits for investing in DESS include:

1) Defer the distribution network upgrade or expansion. Distribution network expansion is usually required because of the load or DG growth. Using DESS to shave the peak load and flatten the daily power profile, the network expansion can be deferred to later years or can even be canceled. This could bring the biggest benefit for the DNO.
2) Price arbitrage. The premise of price arbitrage is that the purchased electricity price is not fixed and changes with time, e.g. the time of use price. Then the DESS can be controlled to charge during the low-price periods and discharge during the high-price periods.
3) Accommodate more distributed generations. References [10] and [11] deal with the optimal allocation of DESS in the distribution system to reduce the wind power curtailment. Except for price arbitrage, to absorb the spilled wind energy can be a benefit for both DG owners and the DNO.
4) Reduce the active power losses. DESS can be used for both active power and reactive power control. By means of proper placement and operation of energy storage systems, the cost of energy losses in distribution systems can be effectively reduced.
5) Improve distribution system reliability. This benefit is specifically analyzed in reference [12], which is quantified based on the comprehensive reliability evaluation considering optimal operation of DESS after contingences. It should be noted that the DESS are required to operate under the islanding mode, which is a technique challenge and requires extra investment.

9.3 Mathematical Model for Planning Distributed Energy Storage Systems

9.3.1 Planning Objectives

In this section, it is assumed that a DNO invests in DESS for price arbitrage and improving the operational performance of a distribution network with DGs. The objective function is as follows:

$$\min \quad F_1 = C_{ES} + C_{LS} + C_{GC} - B_{PA} \tag{9.1}$$

The first item C_{ES} in Eq. (9.1) is the annualized investment and maintenance cost of DESS. The investment cost of a DESS consists of three parts:

- Power-related cost (mainly the cost for the power electronic coverters [14])
- Capacity-related cost (mainly the cost of energy storage units)
- Fix cost (related to land usage, construction costs, etc.).

In reference [10], the battery replacement cost is also considered, which is related to the number of charging/discharging circles in field operation of DESS.

The second item C_{LS} in Eq. (9.1) is the annually active power loss cost. The third item C_{GC} in Eq. (9.1) is the annually total power curtailment loss for DGs including wind and solar power generations. The last item in Eq. (9.1) stands for the benefit of price arbitrage using DESS.

9.3.2 Dealing with Load Variations and Uncertain DG Outputs

The benefits and operational costs incurred by deploying DESS change from one day to another. For a distribution network with DGs, the power outputs of DGs and even loads are uncertain. These factors bring challenge to the optimal planning of DESS. To account for the variations of power absorption and injection, the following strategies have been used:

1) Sequential Monte Carlo simulation [12] or annually chronological data [15].
2) Selecting typical day(s) based on historical data [6, 16].
3) Classifying into a number of load and DG output levels based on their corresponding power distributions [17].

For the first strategy, it is suitable for a heuristic-based or simulation-based decision-making method, while the second strategy can be used for both heuristic-based and mathematical optimization-based methods. For the third strategy, although the long-term statistical characteristics of load, wind, and solar power can be taken into account, the sequential dispatch of DESS charging/discharging cannot be dealt with precisely.

9.3.3 Complete Mathematical Model with Operational and Security Constraints

Here we take a single-stage planning for example. The costs and benefits are all annualized. The operational conditions of a whole year are represented by a number of typical scenarios.

9.3.3.1 Each Item of the Objective Function

It is assumed that S scenarios are considered and each scenario corresponds to a typical day, which stands for $N_{d,s}$ days within a year. The total number of the days that all the scenarios stand for is equal to 365.

1) **Investment and maintenance cost of DESS**

The costs related to build DESS are calculated by the following formulas:

$$C_{ES} = C_{ES,F} + C_{ES,E} + C_{ES,P} + C_{ES,M} \tag{9.2}$$

$$C_{ES,F} = f_F(\boldsymbol{x}_{es}) = p_{es,f} \times \sum_{i \in \psi_{es}} x_{es,i} \tag{9.3}$$

$$C_{ES,E} = f_E(\boldsymbol{y}_{es}) = p_{es,e} \times \sum_{i \in \psi_{es}} y_{es,i} \tag{9.4}$$

$$C_{ES,P} = f_P(\boldsymbol{z}_{es}) = p_{es,p} \times \sum_{i \in \psi_{es}} z_{es,i} \tag{9.5}$$

$$C_{ES,M} = f_M(\boldsymbol{x}_{es}, \boldsymbol{y}_{es}, \boldsymbol{z}_{es}) = \sum_{i \in \psi_{es}} \left(c_{mf} x_{es,i} + c_{me} y_{es,i} + c_{mp} z_{es,i} \right) \tag{9.6}$$

Here ψ_{es} is the candidate node set for installing DESS; $p_{es,f}$, $p_{es,e}$, $p_{es,p}$ are the annualized fixed cost, energy capacity cost and power capacity cost for investing a DESS, respectively; c_{mf}, c_{me}, and c_{mp} are the corresponding annual maintenance costs; $x_{es,i}$ is a binary decision variable, which is equal to 1 when a DESS is built at node i – otherwise, it is equal to 0; and $y_{es,i}$, $z_{es,i}$ are the decision variables of the energy capacity and power capacity for the DESS deploying at node i, respectively.

2) **Total loss cost**

$$C_{LS} = f_{LS}(\boldsymbol{P}_{ls}) = \sum_{s \in \psi_{sc}} \left[N_{d,s} \times \sum_{t=1}^{T} (p_{s,t} P_{ls,s,t} \Delta T) \right] \tag{9.7}$$

where ψ_{sc} is the set of considered scenarios in a whole year, $N_{d,s}$ is the days that scenario s represents, T is the number of intervals taken into account in a day and ΔT is the duration of each interval, and $p_{s,t}$ and $P_{ls,s,t}$ are the electricity purchase price and the total active power loss at interval t under scenario s.

3) **Total loss of DG curtailment**

The curtailment losses of wind and solar power are considered:

$$C_{GC} = f_{GC}(\boldsymbol{P}^{cw}, \boldsymbol{P}^{cpv}) = \sum_{s \in \psi_{sc}} \left[N_{d,s} \times \sum_{t=1}^{T} \left(p_{s,t}^{w} \sum_{i \in \psi_{w}} P_{i,s,t}^{cw} + p_{s,t}^{pv} \sum_{i \in \psi_{pv}} P_{i,s,t}^{cpv} \right) \Delta T \right] \tag{9.8}$$

where ψ_w and ψ_{pv} are sets of nodes where distributed wind turbines and photovoltaic (PV) panels are installed, $p_{s,t}^w$ and $p_{s,t}^{pv}$ are the selling prices of wind power and solar power at interval t under scenario s, respectively, and $P_{i,s,t}^{cw}$ and $P_{i,s,t}^{cpv}$ are the curtailed wind power and solar power of node i at interval t under scenario s.

4) **Total benefit of price arbitrage**

Price arbitrage is calculated by accumulating the electricity purchase cost and sale revenue:

$$C_{PA} = f_{PA}(\boldsymbol{P}^e) = \sum_{s \in \psi_{sc}} \left[N_s^d \sum_{t=1}^{T} \left(p_{s,t}^e P_{s,t}^e \Delta T \right) \right] \tag{9.9}$$

It is assumed that the prices of electricity purchase and sale are the same, i.e. $p_{s,t}^e$ for each node within the same time interval and $P_{s,t}^e$ the power exchange of DESS with the distribution network.

9.3.3.2 Installation Constraints of DESS

$$\sum_{i \in \psi_{es}} x_{es,i} \leq \overline{N}_{es}, \ \forall i \in \psi_{es} \tag{9.10}$$

$$x_{es,i} \underline{P}_{es,i} \leq y_{es,i} \leq x_{es,i} \overline{P}_{es,i}, \ \forall i \in \psi_{es} \tag{9.11}$$

$$x_{es,i} \underline{E}_{es,i} \leq z_{es,i} \leq x_{es,i} \overline{E}_{es,i}, \ \forall i \in \psi_{es} \tag{9.12}$$

Here $\overline{N}_{es}$ means the maximum number of DESS that can be installed, $\underline{P}_{es,i}$ and $\overline{P}_{es,i}$ are the lower and upper limits of DESS power capacity, respectively, and $\underline{E}_{es,i}$ and $\overline{E}_{es,i}$ are the lower and upper bounds of DESS energy capacity, respectively.

9.3.3.3 Operational Constraints for Each Scenario

The operational constraints of scenarios s are as follows:

1) **Power flow constraints**

$$P_{i,s,t}^w + P_{i,s,t}^{pv} + P_{i,s,t}^e - P_{i,s,t}^d$$
$$= V_{i,s,t} \sum_{j:(ij) \in \psi_{bc}} V_{j,s,t} \left[G_{ij} \cos \left(\delta_{ij,s,t} \right) + B_{ij} \sin \left(\delta_{ij,s,t} \right) \right], \forall i \in \psi_n, s \in \psi_{sc}, t \in \psi_t \tag{9.13}$$

$$Q_{i,s,t}^w + Q_{i,s,t}^{pv} + Q_{i,s,t}^e - Q_{i,s,t}^d$$
$$= V_{i,s,t} \sum_{j:(ij) \in \psi_{bc}} V_{j,s,t} \left[G_{ij} \sin \left(\delta_{ij,s,t} \right) - B_{ij} \cos \left(\delta_{ij,s,t} \right) \right], \forall i \in \psi_n, s \in \psi_{sc}, t \in \psi_t \tag{9.14}$$

Here $P_{i,s,t}^{w}$, $P_{i,s,t}^{pv}$, $P_{i,s,t}^{e}$, and $P_{i,s,t}^{d}$ are the wind, photovoltaic, energy storage, and demand active power, respectively, while $Q_{i,s,t}^{w}$, $Q_{i,s,t}^{pv}$, $Q_{i,s,t}^{e}$, and $Q_{i,s,t}^{d}$ are the corresponding reactive powers, $V_{i,\,s,\,t}$ is the voltage magnitude of node I, G_{ij}, B_{ij}, and $\delta_{ij,\,s,\,t}$ are the conductance, susceptance, and voltage phase difference of the branch from node i to j, and ψ_n is the set of nodes excepting the substation node(s), $\psi_t = \{1, 2, ..., T\}$.

2) **Branch power and capacity constraints**

$$P_{ij,s,t} = V_{i,s,t} V_{j,s,t} \left[G_{ij} \cos\left(\delta_{ij,s,t}\right) + B_{ij} \sin\left(\delta_{ij,s,t}\right) \right], \forall (ij) \in \psi_b, s \in \psi_{sc}, t \in \psi_t \tag{9.15}$$

$$Q_{ij,s,t} = V_{i,s,t} V_{j,s,t} \left[G_{ij} \cos\left(\delta_{ij,s,t}\right) - B_{ij} \sin\left(\delta_{ij,s,t}\right) \right], \forall (ij) \in \psi_b, s \in \psi_{sc}, t \in \psi_t \tag{9.16}$$

$$-\overline{S}_{ij}^2 \leq P_{ij,s,t}^2 + Q_{ij,s,t}^2 \leq \overline{S}_{ij}^2, \forall (ij) \in \psi_b, s \in \psi_{sc}, t \in \psi_t \tag{9.17}$$

Here $P_{ij,\,s,\,t}$, $Q_{ij,\,s,\,t}$, and $\overline{S}_{ij}$ are the active power, reactive power, and maximum apparent power of the branch from node i to j, respectively.

3) **Nodal voltage magnitude constraints**

$$\underline{V}_i \leq V_{i,s,t} \leq \overline{V}_i, \forall i \in \psi_n, s \in \psi_{sc}, t \in \psi_t \tag{9.18}$$

$$V_{i,s,t} = V_{i,s,t}^{set}, \forall i \notin \psi_n, s \in \psi_{sc}, t \in \psi_t \tag{9.19}$$

where $\underline{V}_i$ and $\overline{V}_i$ are the lower and upper bounds of the voltage magnitude at node i. Equation (9.19) means that the voltages of the substation node(s) are given.

4) **Output constraints of wind power and photovoltaic power**

Regarding the wind power and photovoltaic power injections, each of them is equal to the maximum generation capability, i.e. $\overline{P}_{s,i,t}^{w}$, $\overline{P}_{s,i,t}^{pv}$, minus the curtailed power:

$$P_{i,s,t}^{w} = \overline{P}_{i,s,t}^{w} - P_{i,s,t}^{cw}, \forall i \in \psi_w, s \in \psi_{sc}, t \in \psi_t \tag{9.20}$$

$$P_{i,s,t}^{pv} = \overline{P}_{i,s,t}^{pv} - P_{i,s,t}^{cpv}, \forall i \in \psi_{pv}, s \in \psi_{sc}, t \in \psi_t \tag{9.21}$$

$$\sqrt{\left(P_{i,s,t}^{w}\right)^2 + \left(Q_{i,s,t}^{w}\right)^2} \leq SR_i^{w}, \forall i \in \psi_w, s \in \psi_{sc}, t \in \psi_t \tag{9.22}$$

$$\sqrt{\left(P_{i,s,t}^{pv}\right)^2 + \left(Q_{i,s,t}^{pv}\right)^2} \leq SR_i^{pv}, \forall i \in \psi_{pv}, s \in \psi_{sc}, t \in \psi_t \tag{9.23}$$

It is assumed in (9.22) and (9.23) that the reactive power outputs of wind and PV can be controlled and confined by the rated capacity of the wind and PV generation SR_i^{w}, SR_i^{pv}, respectively.

5) **DESS-related constraints**

The constraints of DESS include the power injection/absorption and storage energy balance constraints and limits:

$$0 \leq P_{i,s,t}^{e,c} \leq y_i, \forall i \in \psi_{es}, s \in \psi_{sc}, t \in \psi_t \tag{9.24}$$

$$0 \leq P_{i,s,t}^{e,d} \leq y_i, \forall i \in \psi_{es}, s \in \psi_{sc}, t \in \psi_t \tag{9.25}$$

$$P_{i,s,t}^e = P_{i,s,t}^{e,d} - P_{i,s,t}^{e,c}, \forall i \in \psi_{es}, s \in \psi_{sc}, t \in \psi_t \tag{9.26}$$

$$SOC_{i,s,t}^e = SOC_{i,s,t-1}^e + \left(\eta^c P_{i,s,t}^{e,c} - \frac{P_{i,s,t}^{e,d}}{\eta^d} \right) \frac{\Delta T}{z_i}, \forall i \in \psi_{es}, s \in \psi_{sc}, t \in \psi_t \tag{9.27}$$

$$\underline{SOC}_i \leq SOC_{i,s,t}^e \leq \overline{SOC}_i, \forall i \in \psi_{es}, s \in \psi_{sc}, t \in \psi_t \tag{9.28}$$

$$\sqrt{\left(P_{i,s,t}^e\right)^2 + \left(Q_{i,s,t}^e\right)^2} \leq SR_i^e, \forall i \in \psi_{es}, s \in \psi_{sc}, t \in \psi_t \tag{9.29}$$

Here $P_{i,s,t}^{e,c}$ and $P_{i,s,t}^{e,d}$ are the charging and discharging powers of DESS i, respectively and η^c and η^d are the charging and discharging efficiencies, respectively. Regarding the reactive power output $Q_{i,s,t}^e$ of DESS i, it is also limited by the rated power capacity of DESS SR_i^e.

9.4 Solution Methods for the Optimal Distributed Energy Storage System Planning Problem

9.4.1 Second-Order Cone Programming Method

The formulation given in Section 9.3 is a complex mixed integer nonlinear programming problem. It is very difficult to solve it directly. Second-order cone programming (SOCP) formulations have been built for solving some operational optimization problems of distribution networks in recent years, such as the distribution network reconfiguration [18]. The above DESS planning problem can also be solved by the SOCP approach. The nonlinear power flow and network security constraints (9.13) to (9.18) should be reformulated first, as follows [18, 19].

The square of nodal voltages and branch current flows will be used in the constraints:

$$v_{i,s,t} = \left(V_{i,s,t}\right)^2, \forall i \in \psi_n, s \in \psi_{sc}, t \in \psi_t \tag{9.30}$$

$$f_{ij,s,t} = \left(F_{ij,s,t}\right)^2, \forall i \in \psi_n, s \in \psi_{sc}, t \in \psi_t \tag{9.31}$$

Here $F_{ij, s, t}$ is the current of the branch from node i to j.

The power flow equations can be rewritten as

$$P_{i,s,t}^{w} + P_{i,s,t}^{pv} + P_{i,s,t}^{e} - P_{i,s,t}^{d}$$
$$= \sum_{k:(i,k)\in\psi_b} (P_{ik,s,t}) - \sum_{k:(k,i)\in\psi_b} (P_{ki,s,t} - r_{ki}f_{ki,s,t}), \forall i \in \psi_n, s \in \psi_{sc}, t \in \psi_t$$

(9.32)

$$Q_{i,s,t}^{w} + Q_{i,s,t}^{pv} + Q_{i,s,t}^{e} - Q_{i,s,t}^{d}$$
$$= \sum_{k:(i,k)\in\psi_b} (Q_{ik,s,t}) - \sum_{k:(k,i)\in\psi_b} (Q_{ki,s,t} - x_{ki}f_{ki,s,t}), \forall i \in \psi_n, s \in \psi_{sc}, t \in \psi_t$$

(9.33)

The equipment capacity constraints are defined as follows:

$$f_{ij,s,t} \geq \frac{(P_{ij,s,t})^2 + (Q_{ij,s,t})^2}{v_{i,s,t}}, \forall (ij) \in \psi_b, s \in \psi_{sc}, t \in \psi_t$$

(9.34)

$$f_{ij,s,t} \leq f_{ij}^{max}, \forall (ij) \in \psi_b, s \in \psi_{sc}, t \in \psi_t$$

(9.35)

The nodal voltage magnitude is calculated as follows [20]:

$$v_{j,s,t} = v_{i,s,t} - 2(r_{ij}P_{ij,s,t} + x_{ij}Q_{ij,s,t}) - (r_{ij}^2 + x_{ij}^2)f_{ij,s,t}, \forall (ij) \in \psi_b, s \in \psi_{sc}, t$$
$$\in \psi_t$$

(9.36)

$$\underline{v}_i \leq v_{i,s,t} \leq \bar{v}_i, \forall i \in \psi_n, s \in \psi_{sc}, t \in \psi_t$$

(9.37)

Constraints (9.22), (9.23), and (9.29) are now all in the form of a second-order cone [21]. Constraints (9.34) can be modeled as the following second-order cone constraints:

$$\left\| \begin{matrix} 2P_{ij,s,t} \\ 2Q_{ij,s,t} \\ f_{ij,s,t} - v_{i,s,t} \end{matrix} \right\|_2 \leq f_{ij,s,t} + v_{i,s,t}, \forall (ij) \in \psi_b, s \in \psi_{sc}, t \in \psi_t$$

(9.38)

Note that the total active power loss is also a linear function of $f_{ij,s,t}$; thus the optimal DESS placement problem (9.1) to (9.12) and (9.19) to (9.38) is a mix-integer SOCP problem. It can be solved using the commercial or open source software.

9.4.2 Two-Stage Optimization Method

Although the optimal DESS placement problem can be solved by the SOCP approach, it should be noted that the size of this problem increases dramatically with the increase of both the network scale and the number of scenarios

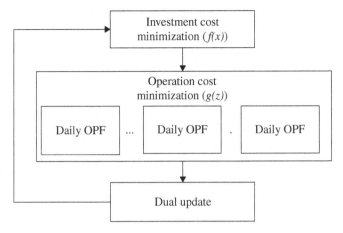

Figure 9.1 ADMM procedure to solve an optimal DESS placement problem.

considered. For a large-scale DESS placement problem, it can be decomposed into two stages or two levels and solved iteratively. In the first stage, a DESS planning decision is obtained, which may be based on engineering experiences, a heuristic method, or a mathematical optimization method. In the second stage, evaluation will be carried out to quantify the concerned benefits of the DESS investment scheme. The results of the second stage will be fed back to the first stage, which can make an updated planning decision using the information. This iterative process will continue until convergence.

If the upper level problem is solved by a heuristic algorithm like the genetic algorithm [6], then the lower-level problem of each scenario is a multi-period OPF problem, as described in Chapter 6. It can be solved by the interior point method or SDP-based method. If a heuristic-based method is used, it will require a large computation burden and the quality of the obtained solution cannot be guaranteed. In reference [22], a decomposition method based on an alternating direction method of multipliers (ADMM) is proposed to solve the large-scale DESS placement problem. The procedure of the ADMM-based method is illustrated in Figure 9.1. Simulation results in reference [22] show that the solution method with ADMM is much faster than the direct solution method without ADMM.

9.4.3 Solution Algorithm Based on Generalized Benders Decomposition

Using reference [23] for the concept and procedure of Benders decomposition, the DESS allocation problem can be decomposed into a storage planning master problem and subproblems reflecting the operational strategy of DESS [24]. To solve the

MISOCP model of Eqs. (9.1) to (9.12) and (9.19) to (9.38) derived above, the generalized Benders decomposition (GBD) approach can be adopted. This approach exploits the structure of the model and reduces the overall computation complexity by decoupling the problem into a master problem and the corresponding subproblems, which can be solved in sequence or parallel. It should be noted that the operation variables in each scenario s are not coupled. Similar to the above-mentioned two-stage optimization approaches, the second-stage operational problem can be decoupled by scenarios when the first-stage planning decision variables $\mathcal{X}$ are given. Thus, the planning problem in the first stage is exactly the master problem and each operation subproblem corresponding to a scenario yields a subproblem.

Let $\mathcal{Y}_s$ denote all the operation variables in scenario s. Then the objective function can be rewritten as

$$\min_{\mathcal{X}, \mathcal{Y}_s} \boldsymbol{f}^T \mathcal{X} + \sum_{s \in \psi_{sc}} \boldsymbol{g}_s^T \mathcal{Y}_s \tag{9.39}$$

where $\boldsymbol{f}$ and $\boldsymbol{g}_s$ are objective coefficient vectors. The first item $\boldsymbol{f}^T \mathcal{X}$ of (9.39) is actually the C_{ES} in (9.1) and the second item of (9.39) corresponds to the other items in (9.1). Then recall the DESS installation constraints (9.10) to (9.12). As they are only associated with $\mathcal{X}$, hereinafter these linear constraints will be represented as

$$\boldsymbol{C}_I \mathcal{X} \leq 0 \tag{9.40}$$

where $\boldsymbol{C}_I$ represents the coefficient matrix of the constraints for $\mathcal{X}$. As for operation constraints (9.19) to (9.38) in scenario s, they can be reformulated as

$$\|\boldsymbol{A}_{sm}\mathcal{Y}_s + \boldsymbol{b}_{sm}\|_2 \leq \boldsymbol{c}_{sm}^T \mathcal{Y}_s + \boldsymbol{d}_{sm}^T \mathcal{X}, \ \forall m \tag{9.41}$$

where m is the index of the second-order cone constraints and $\boldsymbol{A}_{sm}$, $\boldsymbol{b}_{sm}$, $\boldsymbol{c}_{sm}$, and $\boldsymbol{d}_{sm}$ are, respectively, the coefficient matrix and vectors in the second-order cone constraints. Note that $\boldsymbol{A}_{sm} = 0$ means the mth constraint is linear and $\boldsymbol{d}_{sm} = 0$ means it is not associated with $\mathcal{X}$. Therefore, every constraint of (9.19) to (9.38) can be transformed into the second-order cone form. Thus, the optimal DESS planning model has been reformulated as follows:

DESS-P:

$$\min_{\mathcal{X}, \mathcal{Y}_s} \boldsymbol{f}^T \mathcal{X} + \sum_s \boldsymbol{g}_s^T \mathcal{Y}_s$$

$$\text{s.t.} \ \boldsymbol{C}_I \mathcal{X} \leq 0$$

$$\|\boldsymbol{A}_{sm}\mathcal{Y}_s + \boldsymbol{b}_{sm}\|_2 \leq \boldsymbol{c}_{sm}^T \mathcal{Y}_s + \boldsymbol{d}_{sm}^T \mathcal{X}, \ \forall s, \forall m$$

The concepts of the "slave problem (SP)", "dual of SP (DSP)", "master problem (MP)," and "relaxed MP (RMP)" appeared in the Benders decomposition approach and will be defined hereinafter in an orderly way. To begin with, given a first-stage investment decision $\hat{\mathcal{X}}$, the operation SP of scenario s is defined as follows:

SP-s:

$$\min_{\mathcal{Y}_s} \; \boldsymbol{g}_s^T \mathcal{Y}_s$$

s.t. $\|\boldsymbol{A}_{sm}\mathcal{Y}_s + \boldsymbol{b}_{sm}\|_2 \leq \boldsymbol{c}_{sm}^T \mathcal{Y}_s + \boldsymbol{d}_{sm}^T \hat{\mathcal{X}}, \; \forall m$

To generate the 'Benders cut' (mentioned hereinafter), the dual of the convex SP-s should be defined as follows:

DSP-s:

$$\max_{\boldsymbol{\mu}_{sm}, \lambda_{sm}} \; \sum_m \left(\boldsymbol{\mu}_{sm}^T \boldsymbol{b}_{sm} - \lambda_{sm} \boldsymbol{d}_{sm}^T \hat{\mathcal{X}} \right) \tag{9.42}$$

$$\text{s.t.} \; \sum_m \left(\boldsymbol{A}_{sm}^T \boldsymbol{\mu}_{sm} - \lambda_{sm} \boldsymbol{c}_{sm} \right) + \boldsymbol{g}_s = 0 \tag{9.43}$$

$$\|\boldsymbol{\mu}_{sm}\|_2 \leq \lambda_{sm}, \forall m \tag{9.44}$$

Here $\boldsymbol{\mu}_{sm}$ and λ_{sm} are the dual vector and scalar multipliers, respectively. Then the MP can be defined by the following theorem [25].

Theorem 9.1 The optimization problem DESS-P is equivalent to the following problem (named MP):

MP:

$$\min_{\mathcal{X}, \rho} \; \boldsymbol{f}^T \mathcal{X} + \rho \tag{9.45}$$

$$\text{s.t.} \boldsymbol{C}_I \mathcal{X} \leq 0 \tag{9.46}$$

$$\rho \geq \sum_s \sum_m \left(\boldsymbol{\mu}_{sm}^T \boldsymbol{b}_{sm} - \lambda_{sm} \boldsymbol{d}_{sm}^T \mathcal{X} \right), \; \forall (\boldsymbol{u}_s, \Lambda_s) \in \Xi_s, \forall s \tag{9.47}$$

Here Ξ_s is the set of extreme points in the feasible set of DSP. For simplicity, $(\boldsymbol{u}_s, \Lambda_s)$ is used to represent all dual optimization variable of second-order constraints in scenario s, i.e. $(\boldsymbol{u}_s, \Lambda_s) = \cup_m (\boldsymbol{\mu}_{sm}, \lambda_{sm})$.

Then we proof the above theorem. First, as the SP is convex and satisfies the Slater's condition [21] in this case, the strong duality holds:

$$\min_{\mathcal{Y}_s} \; \boldsymbol{g}_s^T \mathcal{Y}_s = \max_{\boldsymbol{\mu}_{sm}, \lambda_{sm}} \; \sum_m \left(\boldsymbol{\mu}_{sm}^T \boldsymbol{b}_{sm} - \lambda_{sm} \boldsymbol{d}_{sm}^T \hat{\mathcal{X}} \right) \tag{9.48}$$

Thus, another equivalent formulation of DESS-P can be derived (the given $\hat{\mathcal{X}}$ in Eq. (9.48) is replaced by $\mathcal{X}$ in Eq. (9.49) because the given planning solution $\hat{\mathcal{X}}$

does not change the feasible set of the DSP, which can be implied in (9.43) and (9.44):

$$\min_{\mathcal{X}} \left(f^T \mathcal{X} + \max_{\mu_{sm}, \lambda_{sm}} \sum_{s \in \psi_{sc}} \sum_{m} (\mu_{sm}^T b_{sm} - \lambda_{sm} d_{sm}^T \mathcal{X}) \right) \tag{9.49}$$

$$\text{s.t.} \, C_I \mathcal{X} \leq 0 \tag{9.50}$$

$$(u_s, \Lambda_s) = \bigcup_m (\mu_{sm}, \lambda_{sm}) \in \Xi_s, \forall s \tag{9.51}$$

Constraint (9.51) indicates that the optimal solution to the DSP must be an extreme point in its feasible set as the DSP is convex. Here, an ancillary variable is introduced:

$$\rho = \max_{\mu_{sm}, \lambda_{sm}} \sum_{s \in \psi_{sc}} \sum_{m} (\mu_{sm}^T b_{sm} - \lambda_{sm} d_{sm}^T \mathcal{X})$$

With variable ρ, the model (9.49) to (9.51) can be transformed to the MP shown above. Thus, Theorem 9.1 is proved.

Note that the constraints set (9.47) actually enumerates constraints for all extreme points in Ξ_s, some of which are useless in obtaining optimality and make the model computationally exhausting. To handle this problem, the GBD approach first relaxes constraints (9.47) and obtains the following RMP:

RMP:

$$\min_{\mathcal{X}, \rho} f^T \mathcal{X} + \rho \tag{9.52}$$

$$\text{s.t.} C_I \mathcal{X} \leq 0 \tag{9.53}$$

$$\rho \geq \sum_{s \in \psi_{sc}} \sum_{m} (\hat{\mu}_{sm}^T b_{sm} - \hat{\lambda}_{sm} d_{sm}^T \mathcal{X}), \, \kappa = 1, 2, \ldots \tag{9.54}$$

Constraints (9.54), namely the "Benders cut", are the set of constraints to be added with each iteration and κ is the index of iterations. $\hat{\mu}_{sm}, \hat{\lambda}_{sm}$, are the optimal solutions of the DSPs in iteration κ.

Then the GBD approach solves the RMP in the first stage and the corresponding DSPs in the second stage iteratively. The lower bound of the objective function comes from the solution of RMP as it relaxes some constraints and the upper bound comes from the solutions of SPs where the investment variables are fixed. In each iteration κ, a "Benders cut" is added to the RMP to force the solution of the RMP to converge to the solution of the original DESS-P problem. The iteration ends when the convergence criterion (relative gap between the upper and lower bounds) is met and the obtained result is guaranteed to be a globally optimal solution [26]. The relaxation of (9.47) and solving the problems with a smaller scale iteratively are the main merits of the GBD approach.

9.5 Distribution Network Expansion Planning with Distributed Energy Storage System

Apart from the deployment of DESS into an existing distribution network that will benefit the optimal network operation, DESS can also be installed at the same time with distribution network planning (DNP). The classical DNP problem has been extensively studied that tries to find the most economical solution with the optimal site and size of extra substations (transformers) and/or feeders to meet the forecasted demands. It should be noted that although the load profile is modeled in DNP as (1) one load level, (2) multi-load levels, (3) probabilistic, or (4) fuzzy [27], a daily load profile is usually not taken into account. However, it is necessary to consider the load variation within a day when DESS placements are simultaneously considered. This will make the DNP problem more complicated.

A complete model for the DNP with DESS will be very complex with many parameters and variables. For simplicity, we show a single-stage formulation and only consider building new branches (feeders) and new DESSs. The replacement of existing feeders and adding new transformers are not taken into account. We also assume that the loads can be fully supplied, i.e. with no load shedding.

With regards to the objective function, the annual total costs should be minimized:

$$\min \quad F_2 = F_1 + C_{\text{DNP}} \tag{9.55}$$

Here F_1 is the same as in (9.1) and C_{DNP} includes the following components:

$$C_{\text{DNP}} = CI_{\text{FD}} + CM_{\text{FD}} \tag{9.56}$$

where CI_{FD} stands for annualized investment costs on feeders and CM_{FD} is the corresponding maintenance cost of newly added feeders:

$$CI_{\text{FD}} = RR_{\text{fd}} \sum_{k \in K_b} \sum_{(ij) \in \psi_{bc}} CI_{\text{fd}}^k \ell_{ij} x_{ij}^k \tag{9.57}$$

$$CM_{\text{FD}} = \sum_{k \in K_b} \sum_{(ij) \in \psi_{bc}} CM_{\text{fd}}^k \ell_{ij} x_{ij}^k \tag{9.58}$$

where K_b is the set of feeder types, ψ_{bc} is the set of candidate feeders, ℓ_{ij} is the length of the feeder from node i to j, CI_{fd}^k and CM_{fd}^k stand for the investment and maintenance costs of a type k feeder per kilometer, respectively, and x_{ij}^k is the investment decision variable. RR_{fd} is the capital recovery rate, which can be calculated by

$$RR_{\text{fd}} = \frac{\alpha(1 + \alpha)^{\eta^l}}{(1 + \alpha)^{\eta^l} - 1} \tag{9.59}$$

where α is the discount rate and η^l is the lifetime of a feeder.

In the following, we will list the new constraints while trying to keep the whole formulation as an SOCP model, given in Section 9.4. The constraints of (9.30) to (9.34), (9.37), and (9.38) can be used. For the candidate feeders, constraints (9.35) should be changed to consider the decision variable as follows:

$$f_{ij,s,t} \le \sum_{k \in K_b} x_{ij}^k f_{ij}^{\max}, \quad \forall (ij) \in \psi_{bc} \tag{9.60}$$

The logic constraints for building new distribution feeders are as follows:

$$\sum_{k \in K_b} x_{ij}^k \le 1 \tag{9.61}$$

Constraints (9.36) should also be modified. To keep linearity, a big M is introduced:

$$\left| v_{j,s,t} - v_{i,s,t} + 2 \left(r_{ij} P_{ij,s,t} + x_{ij} Q_{ij,s,t} \right) - \left(r_{ij}^2 + x_{ij}^2 \right) f_{ij,s,t} \right| \le M \left(1 - \sum_{k \in K^i} x_{ij}^k \right) \tag{9.62}$$

Another type of constraint is that the distribution network should be radial. Almost every published paper on DNP have added constraints to keep the radial structure of the distribution network [28].

First, we count the total number of nodes that has been connected:

$$q_i \le \sum_{j:(ij) \in \psi_{bc}} \sum_{k \in K_b} x_{ij}^k + \sum_{j:(ji) \in \psi_{bc}} \sum_{k \in K_b} x_{ji}^k + m_i \le M q_i \tag{9.63}$$

where m_i is the number of existing feeders that connects to node i and q_i is a binary variable. If a node connects with at least one node, q_i is equal to 1. Otherwise, it is zero. To keep the network radial, the total numbers of connected nodes, substation nodes, and feeders should satisfy the equation

$$\sum_{i \in \psi_n} q_i - N_{ss} = \sum_{i \in \psi_n} \sum_{j:(ij) \in \psi_{bc}} \sum_{k \in K_b} x_{ij}^k + N_{be} \tag{9.64}$$

where N_{ss} is the total number of substation nodes and N_{be} is the total number of existing branches. The first item on the right side of Eq. (9.64) counts the total number of selected candidate branches. It can be seen from Eq. (9.64) that only one line can be built between nodes i and j.

A branch connecting nodes i and j should satisfy the constraints (9.60) to (9.62) and thus the two nodes should have a connection with a substation or a DESS. To eliminate the case that a DESS supply one or more load nodes without the support from a substation, the following constraints can be further added:

$$\sum_{i \in \psi_{ss}} \sum_{k \in K_b} \left(\sum_{j:(ij) \in \psi_b} \tilde{f}_{ij}^k - \sum_{j:(ji) \in \psi_b} \tilde{f}_{ji}^k \right) = N_l \tag{9.65}$$

$$\sum_{k \in K_b} \left(\sum_{j:(ij) \in \psi_b} \tilde{f}_{ij}^k - \sum_{j:(ji) \in \psi_b} \tilde{f}_{ji}^k \right) = 1, \ i \in \psi_l \setminus \psi_{ss} \tag{9.66}$$

$$\sum_{k \in K_b} \left(\sum_{j:(ij) \in \psi_b} \tilde{f}_{ij}^k - \sum_{j:(ji) \in \psi_b} \tilde{f}_{ji}^k \right) = 0, \ i \notin (\psi_l \cup \psi_{ss}) \tag{9.67}$$

$$\left| \tilde{f}_{ij}^k \right| \leq N_l x_{ij}^k, \quad \forall (ij) \in \psi_{bc} \tag{9.68}$$

$$\left| \tilde{f}_{ij}^k \right| \leq N_l, \quad \forall (ij) \in \psi_{be} \tag{9.69}$$

Here ψ_{ss} and ψ_l are the sets of substation nodes and load nodes, respectively, ψ_{be} represents the set of existing branches, N_l is the total number of nodes connected with the load excluding the substation nodes, and $\tilde{f}_{ij}^k$ is the fictitious current that supplies all the load nodes, each with 1.0 p.u. current. The above constraints mean that each load has a path to one substation node, which means the load cannot be supplied by DESS alone. Constraint (9.65) indicates that the total fictitious current supplied from all the substation nodes is equal to the number of loads. Constraint (9.66) stands for the current balance of node loads, while constraint (9.67) sets the current balance of the other nodes. Constraints (9.68) and (9.69) bound the fictitious currents from the candidate and existing feeders, respectively.

It can be seen that the formulation built above is a mixed-integer SOCP model. Although it is more complex than that of only building DESSs, it can be solved by the methods discussed in Sections 9.4.2 and 9.4.3.

9.6 Conclusion and Discussion

The benefits from investing DESS is summarized in this chapter first. The capacity of DESS is much smaller than that of the ESS deployed in transmission systems. Therefore, it is more attractive for both the distribution network owner and third-party investor to build DESS in distribution networks. Two types of optimal DESS planning problems are discussed in this chapter, i.e. DESS planning problems with and without a distribution network reinforcement. A detailed mathematical formulation for DESS deployment without network reinforcement is given by considering the integration of DGs in the distribution system. Furthermore, two types of solution methods are described: SOCP and the two-stage optimization method.

For the distribution network expansion planning problem with DESS, it can be formulated as a mixed-integer SOCP problem and thus can be solved by off-the-shelf software.

References

1 Puttgen, H.B., MacGrego, P.R., and Lambert, F.C. (2003). Distributed generation: semantic hype or the dawn of a new era? *IEEE Power and Energy Magazine* 1 (1): 22–29.

2 Divya, K.C. and Østergaard, J. (2009). Battery energy storage technology for power systems – An overview. *Electric Power Systems Research* 79 (4): 511–520.

3 Haddadian, G., Khalili, N., Khodayar, M., and Shahidepour, M. (2015). Optimal scheduling of distributed battery storage for enhancing the security and the economics of electric power systems with emission constraints. *Electric Power Systems Research* 124: 152–159.

4 Bathurst, G.N. and Strbac, G. (2003). Value of combining energy storage and wind in short-term energy and balancing markets. *Electric Power Systems Research* 67 (1): 1–8.

5 Yuan, Y., Zhang, X., Ju, P. et al. (2012). Applications of battery energy storage system for wind power dispatchability purpose. *Electric Power Systems Research* 93: 54–60.

6 Carpinelli, G., Celli, G., Mocci, S. et al. (2013). Optimal integration of distributed energy storage devices in smart grids. *IEEE Transactions on Smart Grid* 4 (2): 985–995.

7 Sedghi, M., Ahmadian, A., and Aliakbar-Golkar, M. (2015). Optimal storage planning in active distribution network considering uncertainty of wind power distributed generation. *IEEE Transactions on Power Systems*: 1–13.

8 Chakraborty, S., Senjyu, T., Toyama, H. et al. (2009). Determination methodology for optimising the energy storage size for power system. *IET Generation Transmission and Distribution* 3 (11): 987.

9 Nick, M., Hohmann, M., Cherkaoui, R., and Paolone, M. (2012). On the optimal placement of distributed storage systems for voltage control in active distribution networks. *2012 3rd IEEE PES International Conference and Exhibition on Innovative Smart Grid Technologies (ISGT Europe)*. IEEE, pp. 1–6.

10 Atwa, Y.M. and El-Saadany, E.F. (2010). Optimal allocation of ESS in distribution systems with a high penetration of wind energy. *IEEE Transactions on Power Systems* 25 (4): 1815–1822.

11 Xing, H., Cheng, H., Zhang, Y., and Zeng, P. (February 2016). Active distribution network expansion planning integrating dispersed energy storage systems. *IET Generation Transmission and Distribution* 10 (3): 638–644.

12 Awad, A.S.A., El-Fouly, T.H.M., and Salama, M. (2014). Optimal ESS allocation and load shedding for improving distribution system reliability. *IEEE Transactions on Smart Grid* 5 (5): 2339–2349.

13 Zidar, M., Georgilakis, P.S., Hatziargyriou, N.D. et al. (2016). Review of energy storage allocation in power distribution networks: applications, methods and future research. *IET Generation Transmission and Distribution* 10 (3): 645–652.

14 Rufer, A. (November 2017). *Energy Storage: Systems and Components*. Boca Raton, FL: CRC Press.

15 Yang, Y., Li, H., Aichhorn, A. et al. (2014). Sizing strategy of distributed battery storage system with high penetration of photovoltaic for voltage regulation and peak load shaving. *IEEE Transactions on Smart Grid* 5 (2): 982–991.

16 Nagarajan, A. and Ayyanar, R. (2015). Design and strategy for the deployment of energy storage systems in a distribution feeder with penetration of renewable resources. *IEEE Transactions on Sustainable Energy* 6 (3): 1085–1092.

17 Awad, A.S.A., El-Fouly, T.H.M., and Salama, M. (2015). Optimal ESS allocation for load management application. *IEEE Transactions on Power Systems* 30 (1): 327–336.

18 Alnaser, S.W. and Ochoa, L.F. (2016). Optimal sizing and control of energy storage in wind power-rich distribution networks. *IEEE Transactions on Power Systems* 31 (3): 2004–2013.

19 Taylor, O.A. and Hover, F.S. (August 2012). Convex models of distribution system reconfiguration. *IEEE Transactions on Power Apparatus and Systems* 27 (3): 1407–1413.

20 Nick, M., Cherkaoui, R., and Paolone, M. (2014). Optimal allocation of dispersed energy storage systems in active distribution networks for energy balance and grid support. *IEEE Transactions on Power Systems* 29 (5): 2300–2310.

21 Gan, L., Li, N., Topcu, U., and Low, S.H. Exact convex relaxation of optimal power flow in radial networks. *IEEE Transactions on Automatic Control* 60 (1): 72–87.

22 Taylor, J.A. (2015). *Convex Optimization of Power Systems*. Cambridge University Press.

23 Nick, M., Cherkaoui, R., and Paolone, M. (2015). Optimal siting and sizing of distributed energy storage system via alternating direction method of multipliers. *International Journal of Electrical Power and Energy Systems* 72: 33–39.

24 Conejo, A.J., Castillo, E., Minguez, R. et al. (2006). *Decomposition Techniques in Mathematical Programming: Engineering and Science Applications*. Springer Science and Business Media.

25 Fortenbacher, P., Ulbig, A., and Andersson, G. (2018). Optimal placement and sizing of distributed battery storage in low voltage grids using receding horizon control strategies. *IEEE Transactions on Power Systems* 33 (3): 2383–2394.

26 Lin, Z., Hu, Z., Zhang, H. et al. (2019). Optimal ESS allocation in distribution network using accelerated generalized Benders decomposition. *IET Generation Transmission and Distribution*.

27 McDaniel, D. and Devine, M. (1977). A modified Benders' partitioning algorithm for mixed integer programming. *Management Science* 24 (3): 312–319.

28 Lavorato, M., Franco, J.F., Rider, M.J., and Romero, R. (February 2012). Imposing radiality constraints in distribution system optimization problems. *IEEE Transactions on Power Systems* 27 (1): 172–180.

Index

Energy Storage for Power System Planning and Operation, First Edition. Zechun Hu.
© 2020 John Wiley & Sons Singapore Pte. Ltd.
Published 2020 by John Wiley & Sons Singapore Pte. Ltd.